NEW FRONTIERS OF SPACE

BY JEFFREY KLUGER, MICHAEL D. LEMONICK AND TIME CONTRIBUTORS

NEW FRONTIERS OF SPACE

EDITOR Stephen Koepp
DESIGNERS Anne-Michelle Gallero, Skye Gurney
PHOTO EDITOR Patricia Cadley
WRITERS David Bjerklie, Andrew Chaikin, Daniel Cray, Jeffrey Kluger, Michael D. Lemonick, Alex Perry
COPY CHIEF Kathleen A. Kelly
REPORTER Molly Martin
EDITORIAL PRODUCTION David Sloan, Lionel P. Vargas
GRAPHICS EDITOR Lon Tweeten

TIME HOME ENTERTAINMENT
PUBLISHER Jim Childs
VICE PRESIDENT, BRAND AND DIGITAL STRATEGY Steven Sandonato
EXECUTIVE DIRECTOR, MARKETING SERVICES Carol Pittard
EXECUTIVE DIRECTOR, RETAIL AND SPECIAL SALES Tom Mifsud
EXECUTIVE PUBLISHING DIRECTOR Joy Butts
DIRECTOR, BOOKAZINE DEVELOPMENT AND MARKETING Laura Adam
FINANCE DIRECTOR Glenn Buonocore
ASSOCIATE PUBLISHING DIRECTOR Megan Pearlman
ASSISTANT GENERAL COUNSEL Helen Wan
ASSISTANT DIRECTOR SPECIAL SALES Ilene Schreider
DESIGN AND PREPRESS MANAGER Anne-Michelle Gallero
BRAND MANAGER Michela Wilde
ASSOCIATE PREPRESS MANAGER Alex Voznesenskiy
ASSOCIATE BRAND MANAGER Isata Yansaneh
ASSOCIATE BOOK PRODUCTION MANAGER Kimberly Marshall

EDITORIAL DIRECTOR Stephen Koepp

SPECIAL THANKS TO: Christine Austin, Katherine Barnet, Jeremy Biloon, Stephanie Braga, Susan Chodakiewicz, Rose Cirrincione, Jacqueline Fitzgerald, Christine Font, Jenna Goldberg, Hillary Hirsch, David Kahn, Amy Mangus, Amy Migliaccio, Nina Mistry, Stanley Moyse, Dave Rozzelle, Adriana Tierno, Vanessa Wu, Time Imaging

ISBN 10: 1-61893-052-4; ISBN 13: 978-1-61893-052-1
Library of Congress Control Number: 2012941237
Printed in China

We welcome your comments and suggestions about TIME Books. Please write to us at:
TIME Books, Attention: Book Editors, P.O. Box 11016, Des Moines, IA 50336-1016
If you would like to order any of our hardcover Collector's Edition books, please call us at 1-800-327-6388, Monday through Friday, 7 a.m. to 8 p.m., or Saturday, 7 a.m. to 6 p.m., Central Time.

EDITOR'S NOTE The extraordinary color images in this book would not be visible to the naked eye. Light from astronomical objects comes in a wide range of wavelengths, such as ultraviolet and infrared, far beyond what is detectable by human vision. Scientists use different techniques to enhance and distinguish colors, sometimes for aesthetic reasons, but often in order to highlight certain features. For these reasons, a black hole might blaze in magenta, a cloud of hydrogen gas might glow green; even vegetation on Earth might appear brick red in order to distinguish it from a rocky background.

Some of the articles in this book were previously published in TIME magazine and LIFE Books.

COVER Photo composite of Earth's moon and the Lagoon Nebula by R. Jay GaBany

IN THE ANDROMEDA *constellation galaxies known collectively as Arp 272 interact to form a cosmic rose—the smaller of the two, UGC 1813, passes through the larger, UGC 1810.*

CONTENTS

THE "SOMBRERO GALAXY" (MESSIER 104) *is one of the largest galaxies at the southern edge of the Virgo cluster of galaxies. Measuring 50,000 light-years across and with a brightness magnitude of 8.98, the elliptical giant sits just beyond the range of unassisted human sight.*

THE DREAM LIVES ON

BY JEFFREY KLUGER

IN 2010, I WAS FLYING HOME FROM THE MIDDLE EAST WITH NEIL Armstrong and Gene Cernan. I was traveling along on a military morale tour that included not just the first and last men on the moon but also Jim Lovell, the commander of Apollo 13. Our travel arrangements were hardly five-star. On this leg of the journey—as on most of them—we were flying aboard a KC-135 tanker plane, outfitted with perhaps 10 seats that would be too narrow even for coach. The astronauts got first dibs on those. Most of the dozens of other people aboard would simply take a spot on a bench against the bulkhead or, as I usually tried to do, sack out atop a pile of cargo.

But on this leg I snagged a seat—shoulder-to-shoulder and elbow-to-elbow with the two lunar legends. I looked to my left: Armstrong. I looked to my right: Cernan. And it occurred to me that this is exactly what it would have felt like 40 years earlier if the three of us were strapping into an Apollo command module on our way to the moon. I closed my eyes and sat very happily with that thought—unconcerned with the fact that I'm a middle-aged man and I was playing rocket ship. And you know what? I still don't care who knows it.

Space has a way of making sweet, goofy dreamers of us all. From babyhood, we're told tales of looking-glass worlds and wizarding schools where everyday rules—including the very laws of physics—don't apply. And then we get older and give up those places and even blush a bit at the power they once had over us. And yet just overhead, just outside the onionskin of our atmosphere, lies a realm that's far larger—and far less plausible—than all those confected places. But it is irrefutably, impossibly real.

Space is all about size—and it's a size that keeps growing. Our galaxy, with its 200 billion to 400 billion stars, was once all we knew. Then we built telescopes that could peer deeper and found 100 billion other galaxies filling the visible universe too. Even that universe—with its 14-billion-light-year reach in all directions—may be just one of who-knows-how-many-other parallel universes. And however much of the universe we see or otherwise detect may be only a small percentage of what's there. The rest could be mysterious dark matter and dark energy.

Space is also about objects, some of which we can actually fathom—black holes, quasars, supernovas, giant nebulae. And it's about concepts we can barely begin to grasp—superstrings and wormholes and M-theory and quantum superposition, all leading to the inevitable question of what it means or, as Stephen Hawking likes to ask: "Why is there something rather than nothing?"

And space, we're increasingly coming to believe, is about biology, too. How could it not be? Water is one of the most common molecules in the cosmos. Amino acids have been found in meteorites. Our own solar system is awash in the hydrocarbons critical to life. What's more, that familiar solar system is hardly the only one around. The Kepler Space Telescope, along with numerous ground-based telescopes, has been discovering bushels of new exoplanets circling other stars, some of them in the not-too-hot, not-too-cold zone around their suns where liquid water is possible. Even those so-called Goldilocks planets may be just a small part of the story, since they would be home only to life as we know it. It's entirely reasonable to assume (indeed, it's unreasonable not to) that with cosmic chemistry combining and recombining in an infinite number of ways, life unimaginably different from our own could arise in other places—life so unfamiliar that we might not even recognize it if we saw it.

TIME magazine has been covering space since our March 10, 1923, issue, when we reported on a just-released movie about time and relativity that was being newly "exhibited at cinema houses." The film, we wrote, "explains Dr. Einstein's theory of how light rays from the stars are bent by the magnetic attraction of the sun as they pass it, and the verification of this theory by astronomers during an eclipse." (Of course, as Einstein knew, it was gravity, not magnetic attraction, that was responsible, which means that either the movie or—alas—a long-ago TIME reporter, got it wrong.)

TIME has stayed on the space beat all along, through the V2 and Sputnik, Gagarin and Glenn, Gemini and Apollo, the shuttles and the space stations—and through all the unmanned probes that have explored the solar system or turned their telescopic eyes to the stars. In this new book, we bring you the latest developments from all parts of the cosmic world, including not just the big-picture stuff that's happening out there, but the people and the space agencies—and, increasingly, the different countries and private companies—that are driving space exploration in the 21st century. We include some of the latest and most dazzling images captured by the Hubble Space Telescope, the Cassini spacecraft orbiting Saturn, the Mercury Messenger probe and more.

There's hard—sometimes mind-bendingly complex—science in all of that. But there's something more, too—something thrilling and transcendent and even oddly simple. On that same 2010 military morale tour, I was standing in the back of an auditorium in Germany only a few feet from Cernan while a video clip about him played for an audience of servicemen and women, most of whom were young enough to be his grandchildren. At one moment, the Cernan of 40 years ago could be seen and heard bounding across the surface of the moon, kicking up slow-motion clouds of dust and chattering happily and a bit breathlessly. The Cernan of today watched silently, then ticked his head at me, signaling me to come closer. I did, and he leaned in and whispered.

"Man, that was fun," he said. He really needed to say no more.

"AMERICA'S CHALLENGE *of today has forged man's destiny of tomorrow," said Apollo 17 astronaut Gene Cernan in 1972. Cernan, the last man on the moon, is seen here taking in the view.*

LIKE SOMETHING *out of a scene from* Star Trek, *NASA's Hyperwall-2, comprising 128 monitors controlled by the supercomputer Columbia, helps scientists visualize high-dimensional data produced by many telescopes. Here a scientist gets an astronaut's view of the moon.*

STELLAR EXHIBIT

The days of peering only at twinkly specks are long gone as advanced telescopes take us deep into space, revealing the breathtaking variety of the universe. Behold the splendor

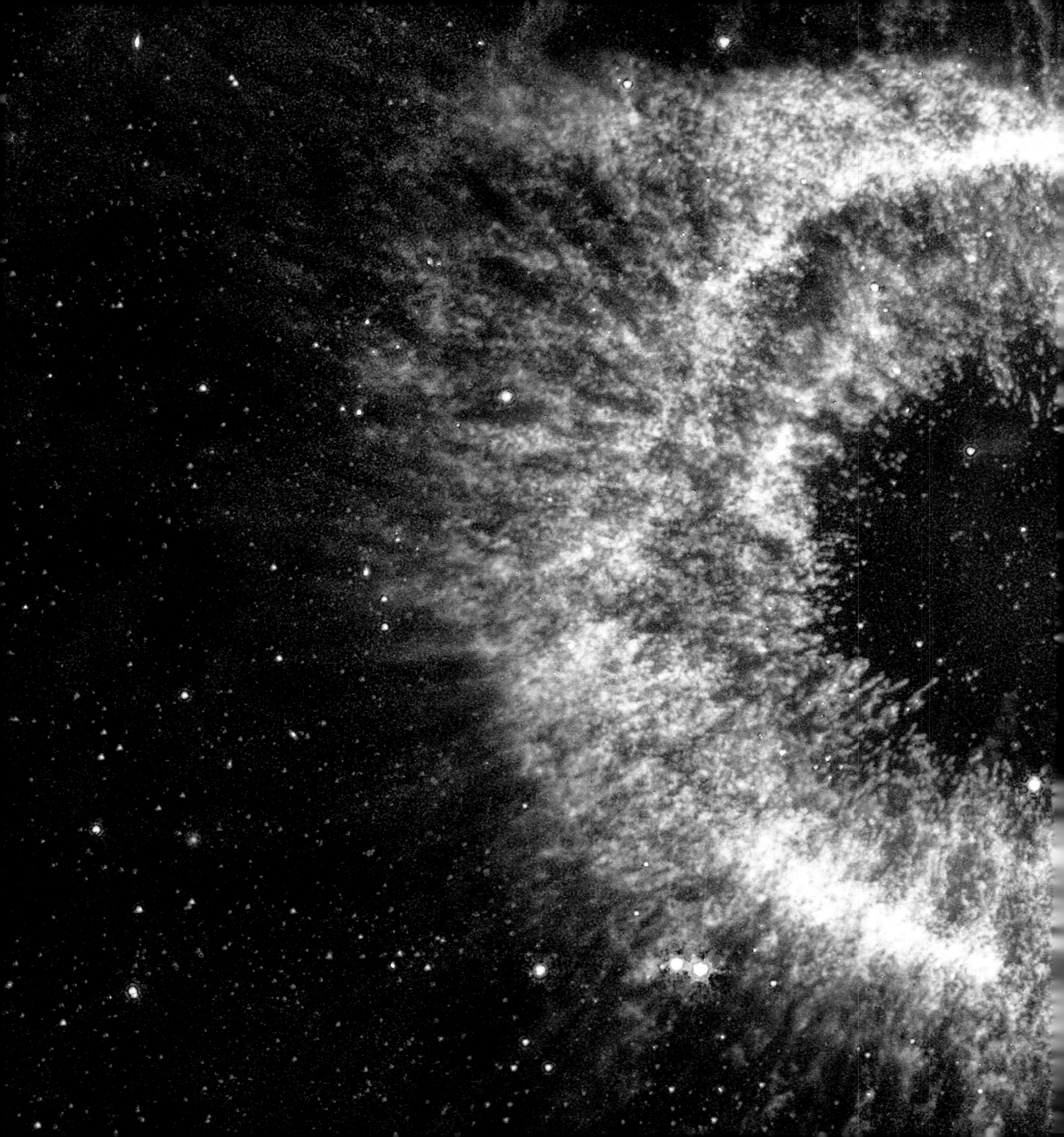

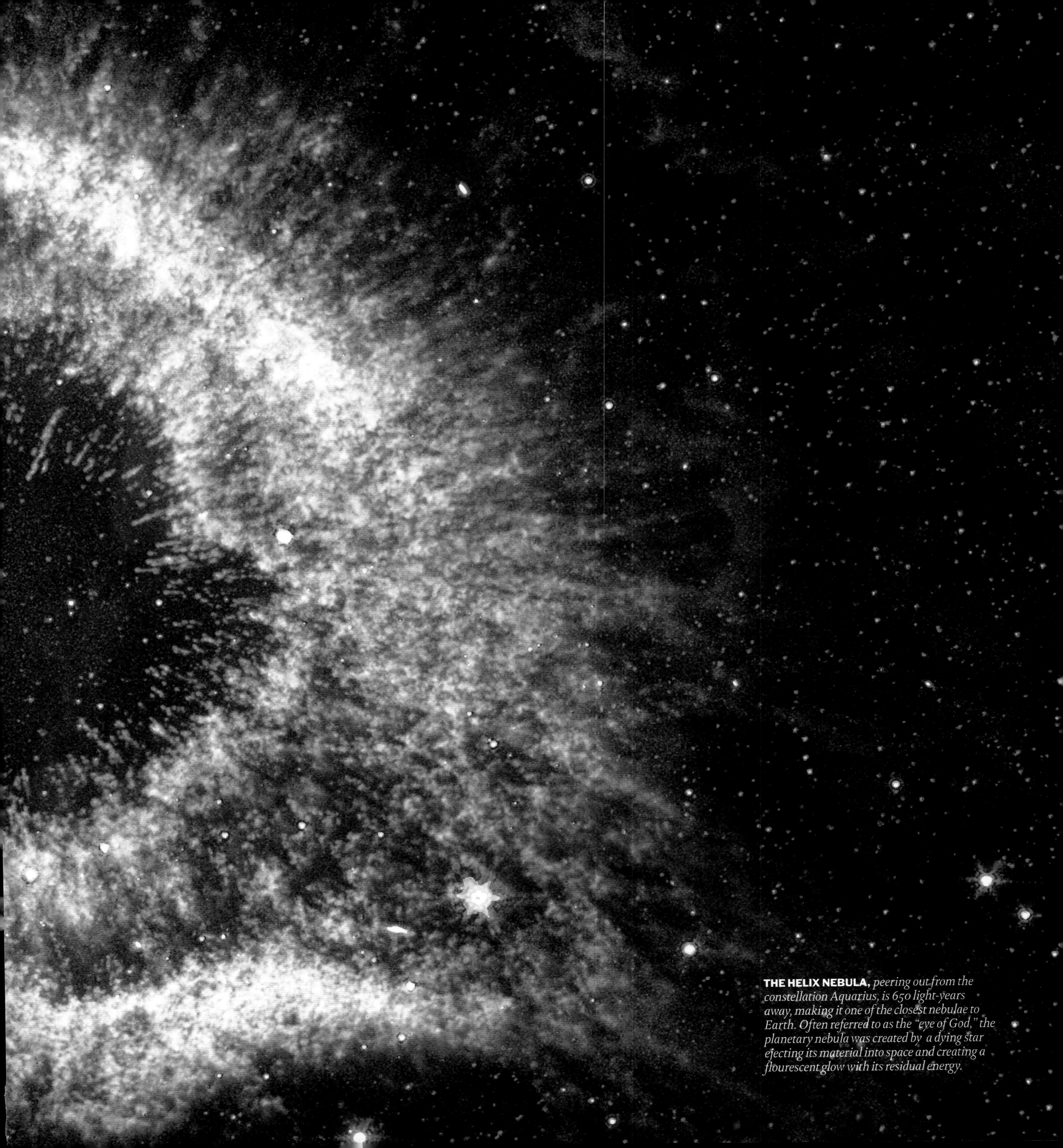

THE HELIX NEBULA, *peering out from the constellation Aquarius, is 650 light-years away, making it one of the closest nebulae to Earth. Often referred to as the "eye of God," the planetary nebula was created by a dying star ejecting its material into space and creating a flourescent glow with its residual energy.*

DUST DEVILS ON MARS *can be 50 times wider than those on Earth. In this case, NASA's Mars Reconnaissance Orbiter captures a 12-mile-high, 70-yard-wide twister along the Amazonis Planitia in 2012.*

A TIME-LAPSE *composite of images taken over six hours by the Solar Dynamic Observatory reveals a rare solar event: the transit of Venus across the face of the sun on June 4, 2012. Venus's previous transit was in 2004, and the next won't occur until 2117.*

JUST AROUND THE CORNER

BY JEFFREY KLUGER

Fleets of high-tech probes and armies of precision telescopes are bringing us fascinating news about our solar system—all of it affirming that there really is no place like home. Not that we know of . . . yet

The solar system, as one Midwestern astrophysicist has put it, is local news. In a universe with billions of galaxies and trillions of stars—never mind the infinite number of parallel universes that, for all we know, exist beyond—we spend all our time in the cosmic equivalent of Hooterville: the same familiar sun; the same eight planets (not even nine anymore); the same old moons and asteroids and odd passing comets. Deep space—where the quarks and quasars and other cosmological exotica live—is far, far beyond us.

But give the local village credit: What it lacks in size, it makes up for in richness. The sun, the planets and the moons especially are forever surprising us with their color, their chemistry, their

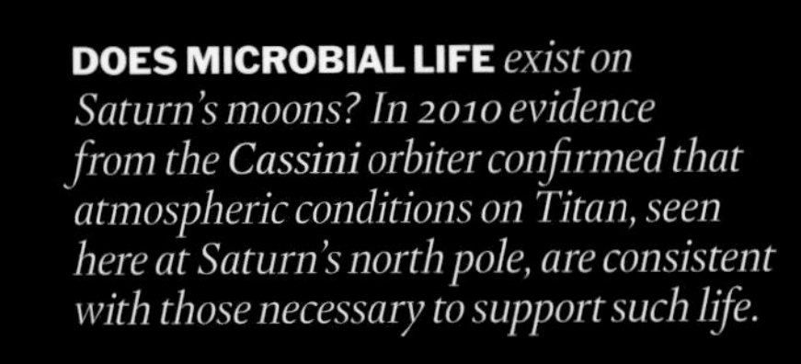

DOES MICROBIAL LIFE *exist on Saturn's moons? In 2010 evidence from the* **Cassini** *orbiter confirmed that atmospheric conditions on Titan, seen here at Saturn's north pole, are consistent with those necessary to support such life.*

complexity—perhaps even their biology.

Little more than half a century ago, humanity had never even gotten a spacecraft as far as the moon. Now we have fleets of probes ranging out all over the solar system, to say nothing of the bristle of ground-based and orbiting telescopes peering out in all directions. What they're discovering reminds us that the old 'hood offers plenty of intrigue after all. If it's not massive new storms roaring up from the sun, it's water on the moon; if it's not the possibility of biology on Saturn's moon Titan, it's quakes on Mercury or rivulets on Mars or a new moon around Pluto or massive oceans on Europa and Enceladus. Everywhere we look, the solar system is serving up breaking news. And while that news may be local, it's thrilling all the same.

The Fiery Parent

The most active part of our very active solar system in the past few years has been the sun itself—and when the sun stirs, everyone feels the effects. Solar activity moves along in its own metabolic rhythms, passing through high and low states of activity in a cycle that lasts about 11 years. Its peak—known as the solar maximum—is characterized by sunspots, magnetic storms, massive flares and bursts of plasma that erupt into space and travel all the way to Earth. The solar minimum is largely free of this turbulence.

The most recent solar minimum, which lasted about twice as long as the two-year average, ended in the summer of 2010—and there was no mistaking when it did. On Aug. 1 of that year, NASA announced "a C-3 class solar flare, a solar tsunami, multiple filaments of magnetism lifting off the stellar surface . . . radio bursts, a coronal-mass ejection." In other words, things got messy.

The mass-ejection part is what most thrills—and terrifies—observers on Earth. On the one hand, the burst of solar energy could disrupt satellite communications all over the planet—a very bad thing, since it's satellites that keep the wired world wired. On the other hand, the charged particles emitted by the coronal-mass ejections spill into the atmosphere at the planet's poles and produce the brilliant sky show known as auroras. The auroras of 2011 occurred as anticipated; the satellite blackout, happily, did not.

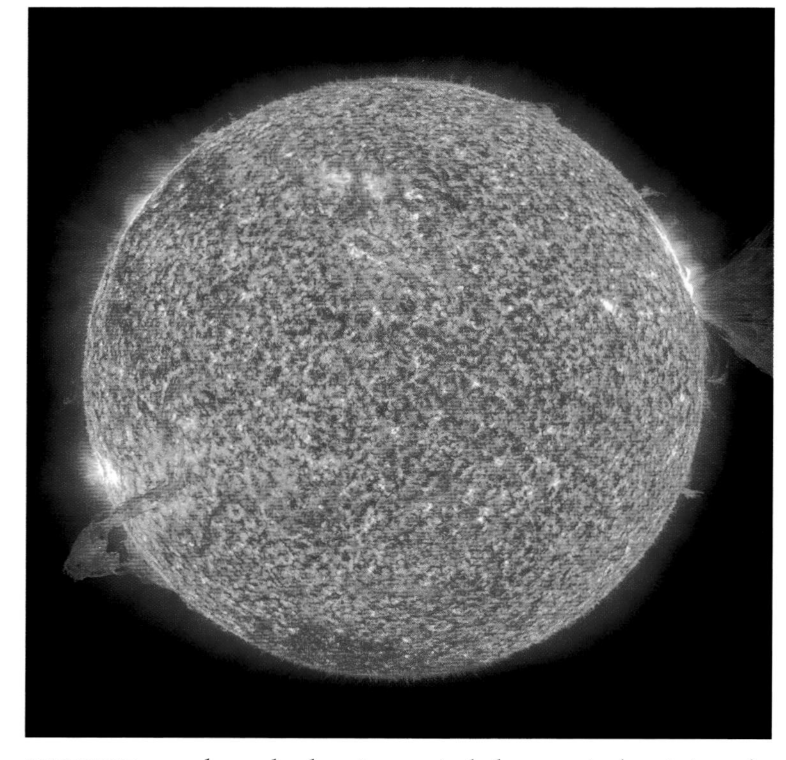

IN RECENT *years the sun has been in a particularly energetic phase in its cycle of activity. Jan. 28, 2011, was an event-filled day: An unstable filament erupted (lower left) while a major flare and a coronal-mass flume ejected (upper right).*

Yet the recent pattern of solar activity may help us predict something with an even more lasting impact on Earth: climate. Solar activity and radiance are thought to account for about 10% of the periodic fluctuations that occur in global temperatures, with solar minimums turning the thermometer down and maximums turning it up. A decades-long minimum in the late 17th and early 18th centuries is thought to have played some role—though probably a small one—in a long stretch of low temperatures known as the Little Ice Age. If cycles are slowing down now—which is what the long minimum that just passed suggests—it's at least theoretically possible that the sun could step in to mitigate some of the effects of humanity's massive emissions of greenhouse gases. That would be nice—though no one has yet proven that those cycles are indeed changing. "There are predictions," says solar astronomer Leon Golub of the Harvard-Smithsonian Center for Astrophysics, "that the coming solar maximum will be a weak one. [But] people are predicting all of the possibilities you could imagine. Somebody's going to end up being right." In other words, don't quit driving low-emission cars anytime soon.

Cosmic Water Park

For all the fire the sun was sending out, most of the news in the solar system lately has been about water—lots of it. First, there's Mars, which astronomers have increasingly understood to have once been a wet—perhaps even verdant—world, like Earth. The planet's surface features certainly suggest that: gullies and channels that resemble riverbeds and deltas; deep depressions that appear to be long-dry oceans and seas. When the Spirit and Opportunity rovers landed on Mars in early 2004, they quickly began discovering salts and other substances that form only in the presence of water.

What was always missing from all those findings was some sign that water currently flows anywhere on the red planet—until 2011, at least. That was when the spacecraft known as the Mars Reconnaissance Orbiter (MRO) spotted seasonal streaking that occurs in a mountainous region of Mars and looks for all the world like water tracks that seep up from the subsurface, flow downslope and then evaporate in the sun.

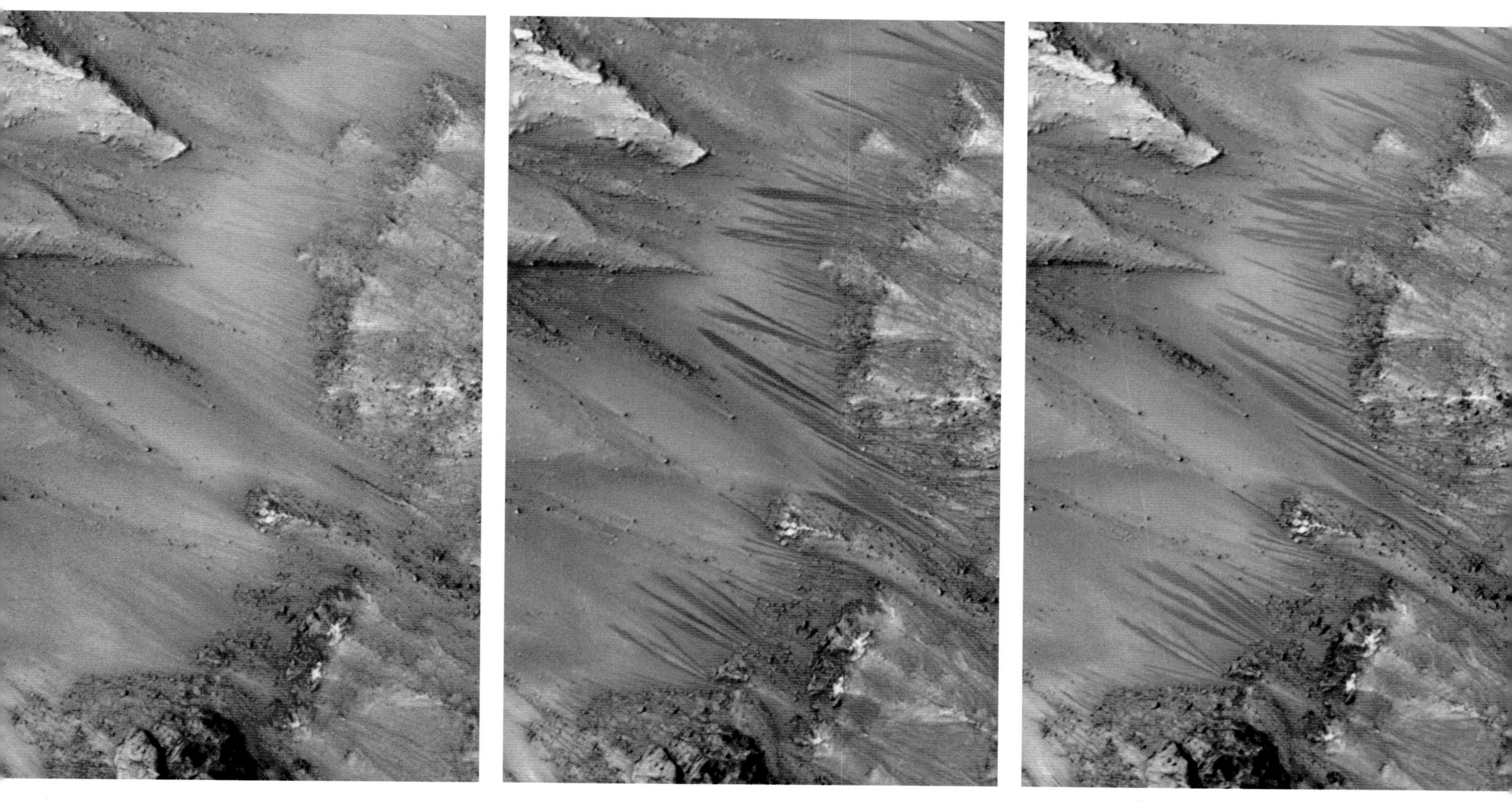

THE NEWTON CRATER ON MARS *strongly suggests that our red neighbor may contain water. This photo series, taken during Martian early spring to midsummer 2011, seems to show seasonal streaking, as if water from beneath the planet's surface is staining the landscape as it evaporates.*

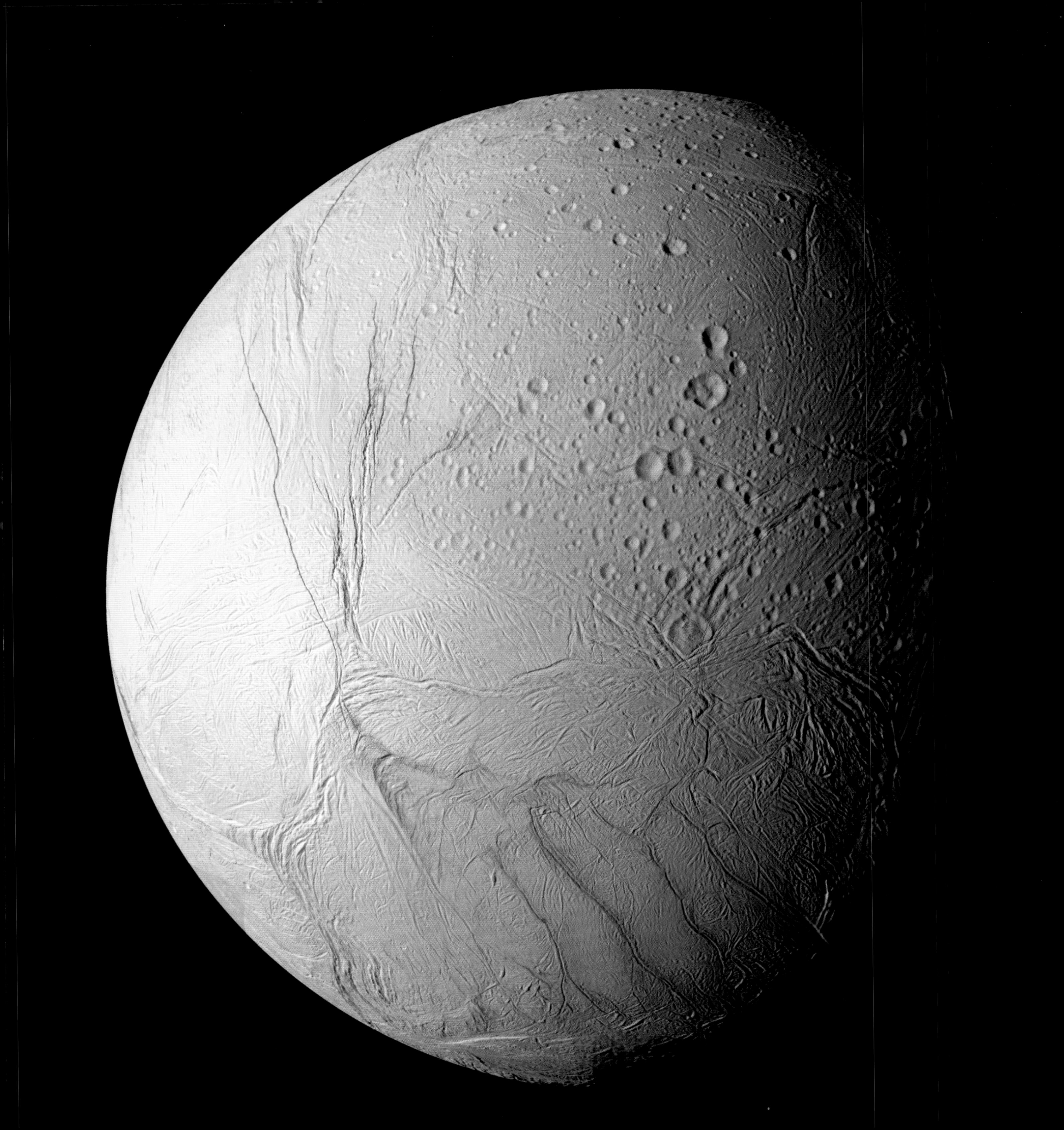

The belief is that these are the product of underground ice deposits that melt in the Martian summer and then retract and solidify in the winter. The seasonality of the phenomenon certainly suggests that, and even the fact that the theoretical water appears to flow when the temperature is a bit below freezing doesn't undermine the theory. Martian water is salty, and salt lowers its freezing temperature. "Since the MRO arrived at Mars, our overarching theme has been 'Follow the water,'" says mission scientist Michael Meyer. "Now we may be catching Mars in the act."

If that's true, it's an act that a number of the solar system's other worlds have also gotten in on. Observations by the European Space Agency's Venus Express orbiter found traces of oxygen, hydrogen and the heavy-hydrogen isotope deuterium in the Venusian atmosphere in precisely the concentrations that would suggest the planet was once rich in water—before it all evaporated in the 800°F heat, that is. "Everything points to there being large amounts of water in the past," says Venus Express science team member Colin Wilson. Jupiter's moon Europa was already assumed to contain a deep—perhaps globe-girdling—ocean of water just beneath its icy rind of a surface. The water is kept liquid by gravitational pulsing that generates heat. Now it turns out that Enceladus, a bright, white Saturnian moon, may be just as waterlogged.

Enceladus's highly reflective surface—relatively free of craters in many areas—suggests that fresh coverings of ice crystals are somehow regularly deposited. Images from the Voyager probes showing a shimmery bulge trailing after Enceladus in one of Saturn's rings suggest that the same kind of gravitational flexing that keeps Europa comparatively warm must also be pumping up Enceladus enough to cause it to emit ice geysers. Much newer images from the Cassini spacecraft orbiting Saturn, just analyzed, reveal wide cracks in the warmer regions of the Enceladan surface, which widen and narrow in cycles corresponding to how much gravitational tugging the moon is experiencing from both its sister moons and Saturn itself as it passes through its orbit. This suggests a world that is more malleable than once thought, and that in turn implies even more subsurface water.

Even our own seemingly desiccated moon is wetter than we ever knew. Scientists had always speculated that water could be preserved in the permanently shadowed regions (PSRs) near the lunar poles, since incoming comets might have deposited ice there, and the deep craters protected by high rims that the impacts produced would shield the ice from the sharply angled sun. The result: Ice could be delivered there and never evaporate.

Various experiments carried out by the Lunar Reconnaissance Orbiter (LRO) had already confirmed that there is some crystalline ice mixed into the moon's regolith, or surface dust. But it's only enough to make the moon about twice as wet as the Sahara desert—though by moon standards that's not bad. New LRO studies, however, have upped this figure, determining that about 2% of the surface area of the PSRs is water ice. What's more, even without sunlight, lunar water exposed to the punishing environment of space would degrade over time, but the fact that so much is still present means that such degradation is happening at about one-sixteenth the speed

WHAT HAS *been dubbed "tiger striping" in these enhanced images of Saturn's Enceladus moon may represent the venting of underground oceans.*

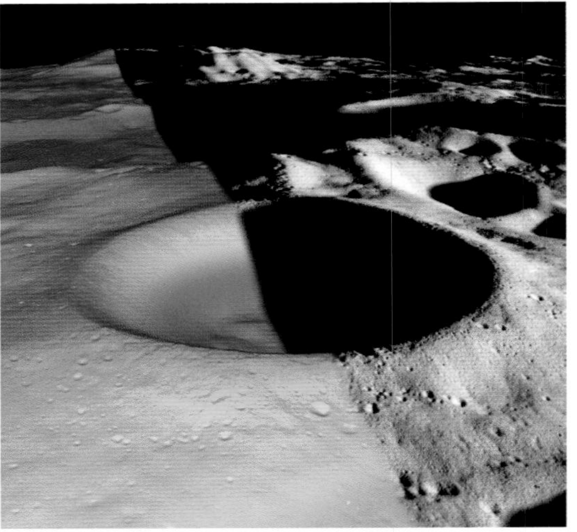

ASTRONOMERS THEORIZE *that water deposited by comets may be preserved within our moon's permanently shadowed deep craters, protected from evaporation and degradation by the high rims created on impact. As much as 22% of the Shakelton crater's surface area (enhanced here using a colorization technique to indicate depth) may be covered in water. It would be a valuable resource for future moon visitors.*

previously estimated. All this is good news, and not just because it makes the moon a more interesting place; it also means that if astronauts ever do set up camps for long-term stays on the moon, much of the water they will need for drinking, growing food and even producing rocket fuel could be harvested on-site rather than lugged up from Earth.

The solar system's modest bookends—little dry-roasted Mercury, closest to the sun, and distant no-longer-a-planet Pluto—have made news, too. Images from the Messenger spacecraft orbiting Mercury reveal uplifted craters and relatively clear, resurfaced areas that point to volcanism—something that was long considered impossible on so dense and structurally simple a world. "There's a lot more evidence for volcanism than we've seen before," said MIT geophysicist Maria Zuber, one of the scientists who made the discovery. "We find some evidence of this dynamism at many places on the surface." And thanks to observations by the closer-to-home Hubble telescope, Pluto was found to have a moon no one knew of before, bringing its litter to four.

Anyone Out There?

The hunt for extraterrestrial life in the solar system has still yielded nothing, though hope remains for something squirreled away beneath the surface of Mars or—who knows?—swimming free in the comparatively warm, briny seas of Europa or Enceladus. But the prospects have brightened a tiny bit for a wholly unexpected type of life on Saturn's moon Titan, which is saturated with the hydrocarbons methane and ethane. Exobiologists have long speculated that organisms might exist on Titan that use liquid methane to perform the same metabolic functions we perform with water. One exceedingly preliminary indicator would be if the atmosphere were low in ethane and acetylene and if hydrogen tended to migrate down to the surface. All of this could theoretically suggest life in the same way rising and falling carbon dioxide levels in our own atmosphere point to the appearance and disappearance of seasonal leaf cover, which essentially inhales CO_2 and exhales oxygen.

In 2010 the Cassini orbiter confirmed that all three atmospheric conditions are met on Titan. But investigators wisely went only so far as to say that those conditions are consistent with—as opposed to evidence for—life. Still, that's a lot better than inconsistent with, and it's something exobiologists longing for the day we find we're not alone would happily accept.

Perhaps the most poignant recent findings about the workings of our solar system concern not the rise of life but the loss of it. In 2012, the centennial of the year the *Titanic* went down, a pair of physicists from Texas State University laid out a new theory suggesting the moon may have been a sort of second perpetrator of the ship's sinking. Early in 1912, the Earth made its closest approach to the sun of the year—a natural part of its regular orbit. But just a day later, the moon made its closest approach to Earth in 1,400 years. The combined gravitational tugs may have raised water levels enough to enable an old iceberg that had calved away from Greenland long ago and gotten hung up in the relatively shallow waters around Newfoundland to float free and drift off—straight into the path of the speeding *Titanic*. That would help explain why the ice was heavier than usual in the shipping lanes of the North Atlantic.

Such a theory, of course, will forever remain just that. Whether or not the distant hands of the moon and the sun did play a role in the disaster, however, the mere possibility is a reminder that what happens in space—especially in our little corner of it—will always have the capacity to touch us. Sometimes that can bring us sorrow. But most of the time it brings us wisdom, insight—and a new appreciation for our place in the cosmos.

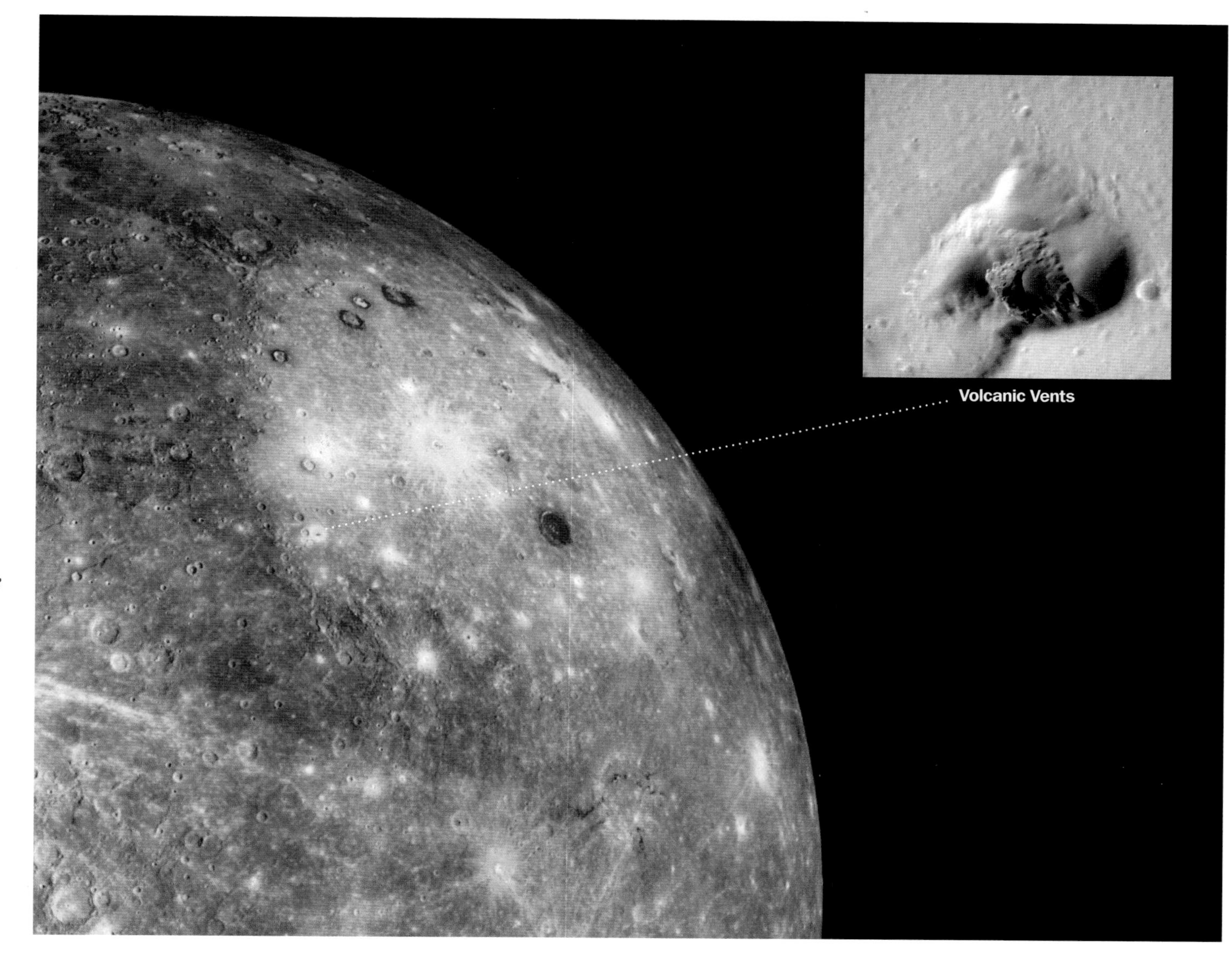

RECENT IMAGES OF MERCURY *have settled an old debate among scientists about the cause of smooth areas within the planet's Caloris basin, one of the solar system's largest impact basins. Orange hues (here enhanced) that dot the Caloris indicate the presence of volcanic vents. The texture and rimless shape of these areas (inset) suggest the existence of lava. The planet may be more geologically active than previously believed.*

How close it came

Asteroid 2012 DA14 passed within 17,200 miles of Earth on Feb. 15, 2013. That's inside the orbit of some satellites

The punch it packed

The 150-ft. rock would have produced a blast 180 times as powerful as the Hiroshima bomb

Other near-Earth objects

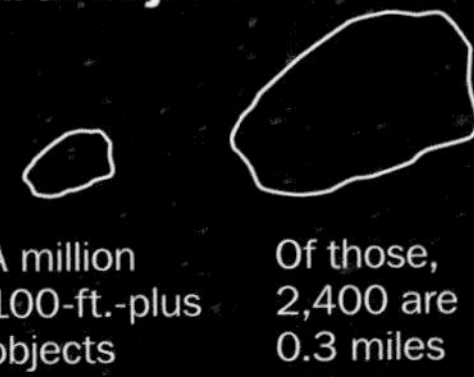

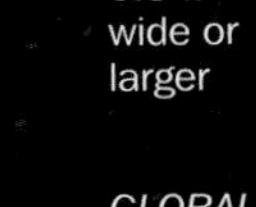

Basketball-size objects hit the atmosphere every day

A million 100-ft.-plus objects are likely out there

Of those, 2,400 are 0.3 miles wide or larger

HARMLESS → *GLOBAL DISRUPTION*

ROCKING THE WORLD

BY JEFFREY KLUGER

Buzzed by two asteroids on the same day, earthlings were reminded of a danger we can't ignore: A direct hit by a big one would be serious trouble

Russia has a right to start feeling picked on. One quiet morning more than a century ago, an asteroid measuring 330 ft. (100 m) across exploded in the skies over the Tunguska region in Siberia, unleashing a blast of up to 30 megatons—or 1,000 times the power of the Hiroshima bomb. Roughly 80 million trees were leveled across an 825 sq. mi. (2,136 sq. km.) expanse. That ought to have been quite enough for any one country—but space, apparently, was not quite through with Russia. On the similarly peaceful morning on Feb. 15, 2013, a smaller meteor exploded over the Ural mountains, injuring 1,500 people and damaging 4,300 buildings over a six-city region.

Remarkably—and scarily—those recent fireworks were only a

INCOMING *The contrail of a meteor streaking across the Russian sky was visible in the dawn light on Feb. 15, 2013; within minutes, hundreds of people were injured and the glass in thousands of buildings was broken. Scientists estimate that 100 tons of space debris enter Earth's atmosphere daily, most of it, fortunately, the size of sand.*

part of the ordnance the cosmos aimed at Earth that day. As NASA had been warning for about a year, an asteroid about the size of a small office building buzzed the planet at the relative tree-top altitude of 17,200 miles (27,359 km) that very evening. If you think that sounds far away, think again; it's actually about 5,000 mi. (8,047 km) below the altitude of our highest-flying satellites. Had that asteroid—known inartfully as 2012 DA14—hit us, it would have meant a 2.4 megaton force, or 180 Hiroshimas. And 2012 DA14, like the one that struck the Urals, has a whole lot of friends.

Astronomers estimate that there are up to 1 million potentially dangerous near-Earth objects (NEOs) out there, only 9,688 of which have been cataloged by astronomers so far. Of those, 1,377 are identified as potentially hazardous asteroids (PHAs), based on their size and their eventual proximity to Earth. And when it comes to space rocks, a word like "hazardous" understates things. Our moon is thought to have been created when a Mars-size planetesimal collided with us more than 4 billion years ago. The dinosaurs were all but certainly wiped out by a 6-mile (10 km) rock that landed off the Yucatán Peninsula about 65 million years back.

"We get 100 tons of interplanetary debris hitting the atmosphere every day," says Don Yeomans, head of the Near Earth Object Program Office at NASA's Jet Propulsion Laboratory. "Most of it is sand size or pea size, but we get a basketball-size object every day. Every few months we get one as big as a Volkswagen."

Even those car-size chunks incinerate before they hit the ground, but the bigger a piece of debris is, the likelier some of it is to survive—and probability says our planet is going to get clobbered again. Unlike the dinosaurs, we have the ability to see it coming and, at least in theory, defend ourselves. We don't have our asteroid shield in place yet, however, and the race is thus on to find—and stop—the next bullet before it finds us.

A Dangerous Neighborhood

Earth, like all the planets and moons in our solar system, has been playing in traffic for a long time. Torturously scarred bodies like Mercury and our moon are testimony to how violent near-Earth space can be, especially going back to the solar system's earlier days, when planets and moons were still accreting and there was plenty of leftover material flying free. Since then, things have quieted down, but only in relative terms.

An asteroid as big as the one that just missed us is estimated to pass close by every 40 years and enter the atmosphere every 1,200 years. Bigger rocks, measuring up to a third of a mile, are less common but would be far more devastating, causing a 5,000-megaton blast and a 7.1-magnitude shock and spreading pain on a continent-wide scale. "We're talking serious trouble here," says Yeomans. Once you get up to 1.3 miles (2 km), growing seasons are disrupted and weather patterns are altered all over the planet.

In the early days, astronomers began requesting funds to track near-Earth objects, and in 1995 Congress authorized NASA to establish a full-time monitoring program. The space agency mostly turned the job over to three observatories—on Maui, in Tucson, and in Socorro, N.M. Those facilities have detected about 98% of all the known NEOs, though 2012 DA14 was spotted by amateur astronomers in Spain, and the WISE infrared spacecraft has found more than 150. There is, says Yeomans, a "small army" of professionals and amateurs worldwide contributing to the work.

Just how you define the *near* in near-Earth is a very precise thing. In order to qualify for the cosmic watch list, an object must come within 1.3 astronomical units (AUs) of the sun. A single AU is the distance from Earth to the sun, or 93 million miles. To earn the sobriquet potentially hazardous asteroid, a rock has to come a whole lot closer—within 0.05 AU, or 4.65 million miles (7.5 million km) of us—and measure at least 330 ft. "The asteroid that just passed is at the threshold of what would cause us worry if it were going to hit us," says Lindley Johnson, director of the Near Earth Object Program Office in NASA's Washington headquarters.

Once any object has been found and logged, its overall risk is calculated on what's known as the Torino Scale. Named after a 1999 conference in Torino, Italy, where it was adopted, the scale is essentially a grid with the probability of impact—from zero to 100%—on its x-axis and the size of the object on the y-axis. It is then ranked on a five-color chart—think the Department of Homeland Security's now defunct terrorism-threat level—going from white (no hazard) through yellow (decorously described as "meeting attention of astronomers") to red (certain collision, capable of causing at least regional devastation of a kind seen only every 10,000 to 100,000 years). The Torino color for 2012 DA14 was a comforting white, since astronomers knew it would miss us, but there is an ever-present risk from any number of bright red rocks to be reckoned with. "We've got a big system that processes this data for us," says NASA research scientist Paul Chodas, who has developed much of the near-Earth-object tracking and analysis software. "There are a lot of asteroids out there."

Earth Strikes Back

A close brush with an asteroid may be scary, but the fact is, it's as good as a distant miss. The rocks that take a direct bead on us are a different matter, and they will have to be wrangled. The truth is that we have no way of protecting ourselves at the moment, but we're getting close.

NASA has been dancing with asteroids and comets for a while. The Dawn spacecraft orbited the asteroid Vesta from 2011 to 2012, then peeled off for its next target, the asteroid Ceres, which it will reach in 2015. In 2001 the NEAR Shoemaker spacecraft touched down on the asteroid Eros. And in 2005 the Deep Impact spacecraft fired a cannonball-like projectile into comet Tempel 1 to blow out a bit of debris and study its composition. All that is good practice for space defense.

The best way to dispatch an incoming asteroid is not to try to blow it up; that could just turn one big rock into a sort of cluster bomb. A safer approach would be to reprise the Tempel 1 impact model, using a collider to change the asteroid's trajectory slightly. It would take a big bullet to move a big rock—one on the order of a couple of metric tons—but spacefaring nations have been throwing heavy loads into the sky for decades. What's more, the farther an asteroid is from Earth when the collision takes place, the greater the difference even a tiny shift in its trajectory would make. "You might only need to alter the object's speed by a few centimeters per second to get it out of the way," says Johnson.

That sounds relatively easy, but it's all still in the drawing-board stage. If 2012 DA14 had been headed straight for us, the single year we had to prepare would have been spent evacuating the likely impact site and hoping our trajectory calculations were correct. "You need 10 to 20 years to design and build and launch," says Yeomans. "If you're at less than 20 years, you're talking civil defense."

For now, the best the U.S. space defenders can do is keep practicing their skills and watching the skies. In 2016 the Osiris-Rex mission will fly to a well-charted near-Earth asteroid, land on it, collect a sample and return home, which should help NASA sharpen its flying skills. The Near Earth Object Office, meantime, is funded at a steady $20 million per year, which is robust enough to keep the telescopes pointed skyward and the observatories staffed. Johnson says he is cautiously optimistic that if an asteroid were spotted today that could strike us within Yeoman's 20-year time frame, we'd probably have the wherewithal to knock it out—provided we hustle. But probably isn't definitely. When the fate of the planet is at stake, we're going to have to do a better job.

COSMIC BULLET *A neat little hole about 30 ft. across was punched in the frozen surface of Chebarkul Lake in the Russian Urals by a shard of the meteor that exploded in the sky in February 2013.*

PICTURE PERFECT
The SUV-size rover touched down on Aug. 6, 2012, after a daringly elegant parachute-assisted landing.

LIVE FROM MARS

BY JEFFREY KLUGER

The arrival of Curiosity, the most advanced rover ever, ushers in a new era of planetary exploration. The one-ton robotic explorer will teach us a lot about the red planet—and the blue one

The folks in mission control at NASA's Jet Propulsion Laboratory ate a lot of peanuts in the minutes leading up to the landing of the Curiosity rover on Mars. Ever since July 31, 1964, when the Ranger 7 probe was making its final approach to the moon, peanuts have been the order of the day at JPL when a spacecraft is preparing to land. The Ranger's job was a simple one: to crash-land on the lunar surface, snapping a few thousand pictures on the way down to beam back home. Still, six Rangers before it had failed, and the JPL engineers knew they were about out of chances. Ranger 7 at last broke that losing streak, and as it happened, someone was nibbling peanuts during the landing. That, the missile men of JPL figured, must have been a good-luck charm—and no one has dared to defy it since.

But it would take more than luck and peanuts to get Curiosity safely to the surface of Mars. At 1:25 a.m. E.T. on Aug. 6, 2012, the SUV-size rover, sealed inside a blunt-bottomed capsule, would slam into the Martian atmosphere at a blazing 13,000 mph (21,000 km/h). Seven miles above the surface, when the thin air had slowed the ship to 900 mph (1,450 km/h)., its heat shield would pop away, and it would deploy a parachute. Its retrorockets would then bring the rover and its housing to a near hover just two stories above the surface, where it would be lowered to the ground by wire cables—a $2.5 billion extraterrestrial marionette, settling its wheels gently into the red soil.

In Chicago, the Adler Planetarium held a late-night pajama party so families could follow the landing live. In New York City, crowds gathered in Times Square to watch on a giant screen that usually shows only ads. NASA live-streamed the event, and the traffic was so great—up to 23 million people watching in the four hours immediately surrounding the landing—that the servers crashed.

What the people watching the live feed saw was not a spacecraft approaching Mars but a roomful of controllers in matching blue shirts, muttering about data acquisition and imager activation and drogue deployment and more. While much of it was incomprehensible, it was clear that something good was building. And then flight-dynamics engineer Allen Chen called, "Stand by for sky crane," and the room fell silent. Less than a minute later, he announced, "Touchdown confirmed! We're safe on Mars!" With that, the silence was broken—explosively. "That rocked!" exclaimed deputy project manager Richard Cook at the celebratory press conference that followed. "Seriously, was that cool or what?

It was cool, indeed, but it was much more, too. In an era in which the grind and gridlock of Washington have made citizens wary of anything the government touches, this was a reminder of what the country can still do. The scene in mission control was what smart looks like. It was what vision looks like. Retrorockets could have eased Curiosity straight down to the surface, but that would have stirred up too much dust, perhaps fouling its works before it even got started. So the engineers chose the hard and creative and dangerous solution for the simple reason that it was also the best one.

A country that can't get its roads and bridges fixed at home actually has infrastructure on Mars. Two NASA orbiters—Mars Global Surveyor and Mars Odyssey—helped relay Curiosity's transmissions to Earth and wave their newcoming sister in for her landing. And even as Curiosity settled down to work, no fewer than eight other NASA probes were ranging through the solar system, exploring—or on their way to explore—the moon, Mercury, Jupiter, Saturn, Pluto, the asteroid Ceres and the interstellar void beyond the planets.

It's Curiosity, however, that may be the state of the exploratory art. With 10 instruments weighing a collective 15 times more than those aboard the earlier golf-cart-size rovers Spirit and Opportunity, it is study-

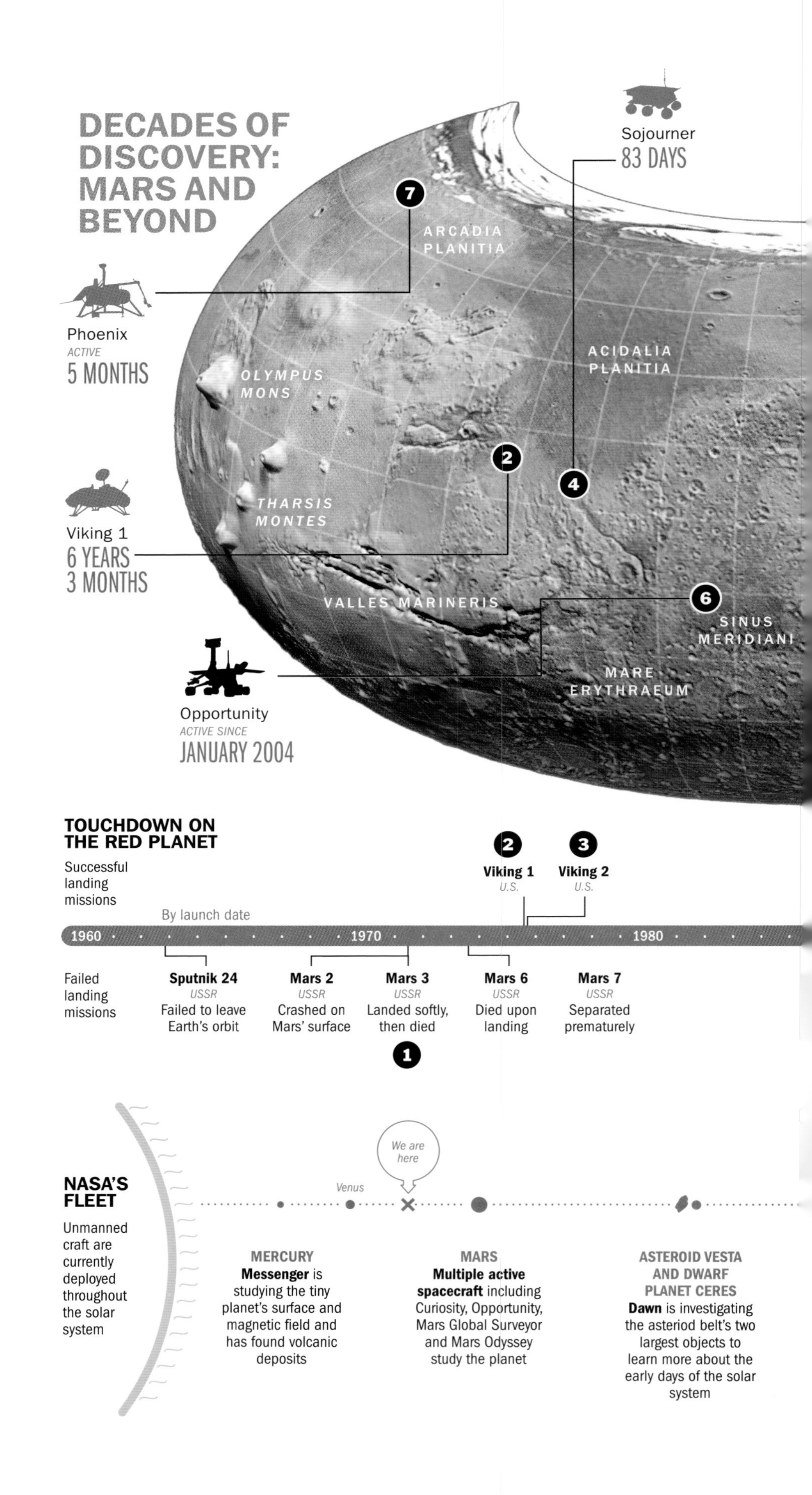

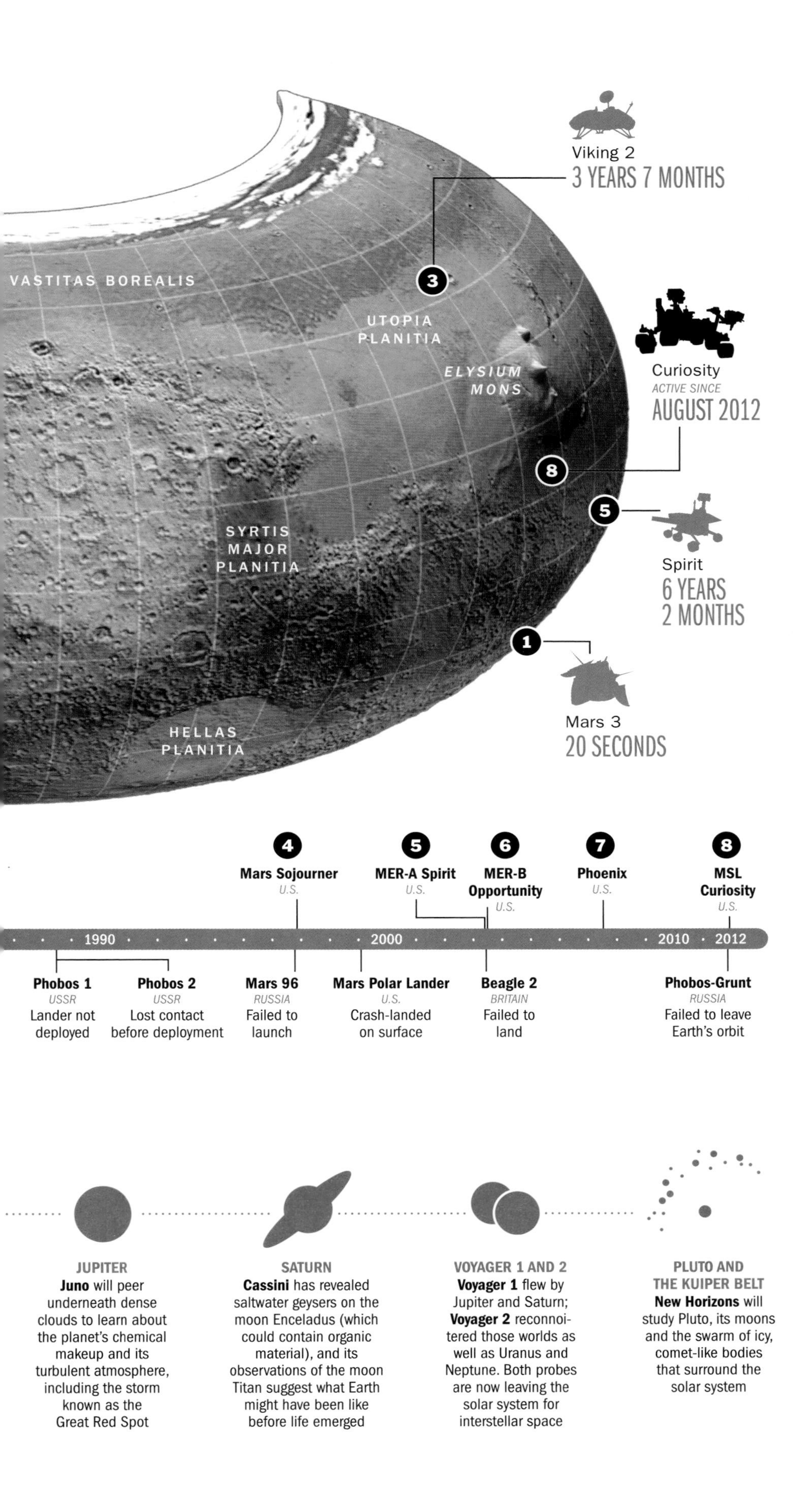

ing the geology, chemistry and possible biology of Mars, looking for signs of carbon, methane and other organic fingerprints on a world that a few billion years ago was warm and fairly sloshing with water. The previous rovers and landers have strongly made the case that Martian life—either extant or ancient—is possible, teeing Curiosity up to seal the deal. "We all feel a sense of pressure to do something profound," says geologist and project scientist John Grotzinger.

In the months since Curiosity landed, it's set about doing just that: photographing its surroundings, sampling the air, trundling over to points of interest and, in early February 2013, using its drill system to collect its first sample of Martian rock and analyzing it in its internal lab for signs of potential biology. All this has framed an unavoidable question: If we can do this exceedingly hard thing so well, why do we make such a hash of the challenges at home, the inventing and investing that 21st-century progress demands? Help answer that one, and Curiosity could achieve great things on two worlds at once.

Follow the Water

Mars may be a meteor-blasted desert today, but it was once a very different place. Its surface is marked with dry riverbeds, empty sea basins and even dusty oceans. Strip away 99% of Earth's atmosphere and boil off all its water and it would look a lot like its desiccated cousin. Mars was wet for at most a billion of its 4.5 billion years, but as the early Earth proved, that could be enough time to cook up life.

As the generation of Mars ships that began flying in the late 1990s discovered, the surface chemistry of Mars is consistent with a once-waterlogged planet. The Spirit and Opportunity rovers used scrapers, drills and abrasion tools to uncover a wealth of minerals that form only or mostly in the presence of water—including salts, gypsum, calcium sulfate and a material known as hematite. The Mars Reconnaissance orbiter found seasonal streaks forming and disappearing on a Martian slope—a sign of underground deposits of existing water that thaw and flow in the Martian spring and freeze and contract in the winter.

Curiosity's landing site is a formation known as Gale Crater, 96 miles wide. Located in the southern Martian hemisphere, it is thought to be up to 3.8 billion years old—well within Mars' likely wet period and thus once a large lake. A 3-mile-high peak known as Mount Sharp rises in its center, with exposed strata layer-caked down its sides. Channels that appear to have been carved by water run down both the crater walls and the mountain base, and an alluvial fan—the radiating channels that define earthly deltas—is stamped into the soil near the prime landing site. All this is irresistible to geologists searching for the basic conditions for life. "We're hoping to find materials that interacted with water," says Grotzinger. Previous landers, he says, did some soil analysis, "but this time we'll find the actual chemicals."

Curiosity is conducting that search in a lot of ways. The rover's arm is designed to scoop up samples of soil and deliver them to an onboard

THE SUCCESSFUL COMPLETION *of a difficult maneuver on Aug. 5, 2012, sent the relieved Spaceflight Operations team into celebration mode. The rover Curiosity contains 10 science instruments for analyzing the geology and chemistry of the Martian environment as it searches for signs of microbial life, past or present.*

analysis chamber, where they are studied by a gas chromatograph, a mass spectrometer and a laser spectrometer, looking for telltale isotopes, gases and elements. Chemical sniffers sample the Martian air for carbon compounds—especially methane—which are the building blocks and by-products of life. Martian geology can be studied with a long-distance laser that can blast a million-watt beam at rocks up to 23 ft. (7 m) away, vaporizing them and allowing a spectrometer to analyze the chemistry of the residue. An onboard X-ray spectrometer can do similar work on rocks near the rover. "With X-ray diffraction, we can really nail down what kind of mineral is there and how those rocks have formed," says deputy project scientist Joy Crisp.

Most appealing for the folks back home are the 17 cameras arrayed around Curiosity. They have the visual acuity to resolve an object the size of a golf ball 27 yd. (24.7 m) away and the resolution to capture one-megapixel color images from multiple perspectives. The sharpest of these imagers is mounted atop the rover's vertical mast, which rises 7 ft. (2.1 m) above ground. "You could not look this thing in the eye unless you were an NBA player," says mission systems manager Mike Watkins. Perhaps the most haunting picture the rover has produced so far, however, was taken from much closer to the ground: a nighttime image of the surrounding terrain, lit only by white and ultraviolet lights from the rover itself.

A Messenger From Earth

The impulse to sentimentalize Curiosity—to treat it almost like a human astronaut—is hard to resist, something that's always been the case with our space machines. "The rover is getting ready to wake up for its first day in a new place," said mission manager Jennifer Trosper at a post-landing news conference. Describing what the science team's schedule would be like, Watkins said, "The rover's day ends on Mars around 3 or 4 p.m. The rover tells us what she did today, and that . . . lets us plan her day tomorrow."

Such anthropomorphizing is the case with space in a way it isn't with other scientific endeavors. The confirmation of the Higgs boson earlier this summer was a much bigger development than the Curiosity landing, but few people—outside the physics community, at least—felt terribly cuddly about it. Nobody calls a particle she.

President Obama—like every President from Kennedy through the second Bush—was quick to make hay out of good news from space. "Tonight, on the planet Mars, the United States of America made history," he said in a post-landing statement. NASA administrator Charles Bolden Jr.—like every agency chief before him—was quick to give props to the President who appointed him. "President Obama has laid out a bold vision for sending humans to Mars in the mid-2030s," he said. "Today's landing marks a significant step."

SPACE TRAVEL AT 140 CHARACTERS

The Curiosity rover is the first spacecraft to have its own Twitter feed: @MarsCuriosity, written by social-media specialists. But that means parody was sure to follow. Here are dueling tweets from the real rover and its tongue-in-cheek twin, @SarcasticRover.

@MarsCuriosity

26 Nov 11 I HAVE LIFTOFF! 28 Dec 11 Are we there yet? 221 days and 296.3 million miles to go till I land on the surface of Mars. **6 Jul 12** Zoom! I'm speeding towards Mars at nearly 48,000 mph relative to the sun. Countdown to landing: 30 days. **6 Aug 12** Entering Mars' atmosphere. 7. Minutes. Of. Terror. Starts. NOW. #MSL **6 Aug 12** Parachute deployed! Velocity 900 mph. Altitude 7 miles. 4 minutes to Mars! #MSL **6 Aug 12** Backshell separation. It's just you & me now, descent stage. Engage all retrorockets! #MSL **6 Aug 12** I'm safely on the surface of Mars. GALE CRATER I AM IN YOU!!! #MSL **6 Aug 12** To the entire team & fans back on Earth, thank you, thank you. Now the adventure begins. Let's dare mighty things together! #MSL

This kind of inspirational talk may be understandable—somebody's got to do the touchdown dance, and the President and NASA chief are the obvious choices. But Obama's record on space has been mixed. The idea of privatizing the business of getting cargo and astronauts to low Earth orbit raised a lot of eyebrows at first, but industry has been quick to move into the commercial space the President opened up, with Elon Musk's SpaceX Corp. and a handful of other companies now queuing up for a lot of paying work flying both manned and unmanned missions. Still, it's hard to say how private that effort has really been so far. NASA has shared some of the R&D costs with its candidate companies and signed lucrative contracts with them before they even proved they were up to the job—to the tune of more than $4 billion covered by taxpayers.

The President's plan gets less clear—and less credible—when it comes to manned travel to deep space. NASA is developing a new crew vehicle called Orion—essentially a souped-up Apollo spacecraft—and a new heavy-lift booster dubbed the Space Launch System (SLS), similar to the venerable Saturn V. Returning to the old model of the expendable booster with the crew vehicle perched on top is a safe and smart decision after the disasters of the shuttle era, but that old model was well funded—and the speed and scale of the accomplishments that followed proved that the money was well spent. The first Saturn V was launched in 1967, the 13th and last in 1973, and nine of those rockets took people to the moon.

The SLS, which in one form or another has been in the planning stage since 2004, is not scheduled for its first unmanned flight until 2017 or its first manned one until 2021. After that, it would fly every other year—at best. It's not even clear what its destination would be—perhaps an asteroid, perhaps Mars, perhaps somewhere else. "This is a pace that doesn't make any sense," says John Logsdon, professor emeritus at George Washington University's Space Policy Institute. "When Kennedy said he'd get to the moon by the end of the decade, he actually meant 1967, and he thought he'd still be President."

Kennedy, of course, wasn't hamstrung by budget issues, and Obama's space team is quick to point that out. "I would be thrilled if we could land on Mars in the 2030s, and I truly believe that is within the capability of this country," says John Grunsfeld, head of the NASA science directorate and a five-time shuttle astronaut. "I don't believe it is necessarily within the capability of this country with a flat budget."

For the unmanned program, a flat budget would actually be an improvement. Funding for Mars missions was set to fall from $587 million in 2012—that's million, with an m—to $360.8 million in 2013, causing the U.S. to drop out of a planned collaboration with the European Space Agency for two missions, one of which would have returned a sample from the Martian surface. Already in the pipeline is a new NASA orbiter that will launch later in 2013, but after that, no missions are scheduled until 2018 and 2020—maybe. Says Grunsfeld: "We can just barely afford those missions."

What any nation can afford, of course, is at least partly a function of what it chooses to afford, even in straitened circumstances. The genius of Kennedy's commitment to a lunar landing before 1970 was its simplicity—a single goal and a deadline. The current plan—with ever changing destinations and dates—has none of that New Frontier clarity.

Kennedy, however, had the help of other Presidents and a cooperative Congress. The space push spanned four administrations—counting that of Eisenhower, who created NASA—and six Congresses. And while they often battled over the budget, they agreed on the goal. A legislature that can barely keep the FAA funded is not an easy partner for any White House with grand ambitions. Space isn't free, but with NASA's budget hovering in the vicinity of just $15 billion per year, or 0.47% of the total federal budget, it's hardly a bank breaker either. The Department of Defense, by contrast, gets $716 billion, or 18.9%. What's more, as with Defense, NASA research pays dividends. The Curiosity program has employed 7,000 high-tech workers in most of the 50 states. And as Grunsfeld points out, the rover's chemical sniffers—sensitive to individual organic molecules—could have national-security applications at ports and airports.

But the extraordinary success of the Mars Curiosity rover masks a far greater truth about space exploration: It requires monomaniacal commitment and an exceedingly high tolerance for failure. In their own way, the JPL peanuts are a reminder of that fact. It's unimaginable in today's attention-deficit political climate that there ever would have been a Ranger 7 after the repeated failures of Rangers 1 through 6. But it took mastering unmanned crash landings before we could master unmanned soft landings. And it took mastering unmanned soft landings before Neil Armstrong—five years almost to the day after Ranger 7 made its suicide plunge into the moon's Sea of Clouds—could set his boot onto the Sea of Tranquility. That's the way science progresses: incrementally, patiently and ultimately spectacularly. Some of America's grandest moments have come when we've trusted that fact.

@SarcasticRover

6 Aug 12 Great ... 100,000,000 miles and I'm stuck in a damn crater. Awesome. **6 Aug 12** Mars can get down to –153 Celsius. Good thing no one bothered to pack me a sweater! Idiots. **6 Aug 12** What they don't show you in that parachute photo is me inside the capsule pissing myself in sheer terror. **6 Aug 12** I'm going to take a nap ... even though all the science here is super important and ... LOL JK, I'M JUST SIFTING DIRT LIKE A BEACH-HOBO **7 Aug 12** Just looked at a rock and it was all "What are you lookin' at?" LOL so I nuclear-lasered it and now the others know who's in charge. **7 Aug 12** MARS isn't just a beige wasteland!! There's also taupe and tan and kindofyellowishbrown!! It's the most colorful planet!! SCIENCE!! **7 Aug 12** YOU GUYS!!! I FOUND A BLACK HOLE! Oh ... no wait, never mind. Just a regular hole. IT'S JUST A REGULAR HOLE. STAND DOWN! Sorry!

TRIUMPH OF THE PLANET HUNTERS

BY MICHAEL D. LEMONICK

New discoveries suggest that our galaxy contains billions of Earth-size worlds. Now astronomers are stepping up the search for those planets that truly resemble our own in the critical ways that could support life

Look up on a clear, moonless night, far from city lights, and you can see thousands of stars twinkling in the dark vault of the sky. Astronomers will assure you that the stars aren't really twinkling, of course. It's an illusion caused by the shimmering of Earth's atmosphere, something like the rippling air you see coming off the hood of a car on a hot summer day.

But in fact, many of those stars *are* twinkling, in a way that only a clear-eyed telescope like the Hubble, orbiting high above the atmosphere, can make out. More precisely, the stars are winking at us, dimming just a tiny bit on a schedule as precise as clockwork. Others are wobbling in place, moving toward us and away and toward us with the same sort of rhythm.

BY MEASURING *the winks and wobbles of light from suns when their planets pass in front of them, astronomers have found thousands of new planets. At right (clockwise from top left), Kepler 11's six planets orbiting their sun; Kepler 22b, the first planet discovered to orbit the "habitable zone," where water could exist; gas giant H189733b passing its star; planet Kepler 16b circling its two stars*

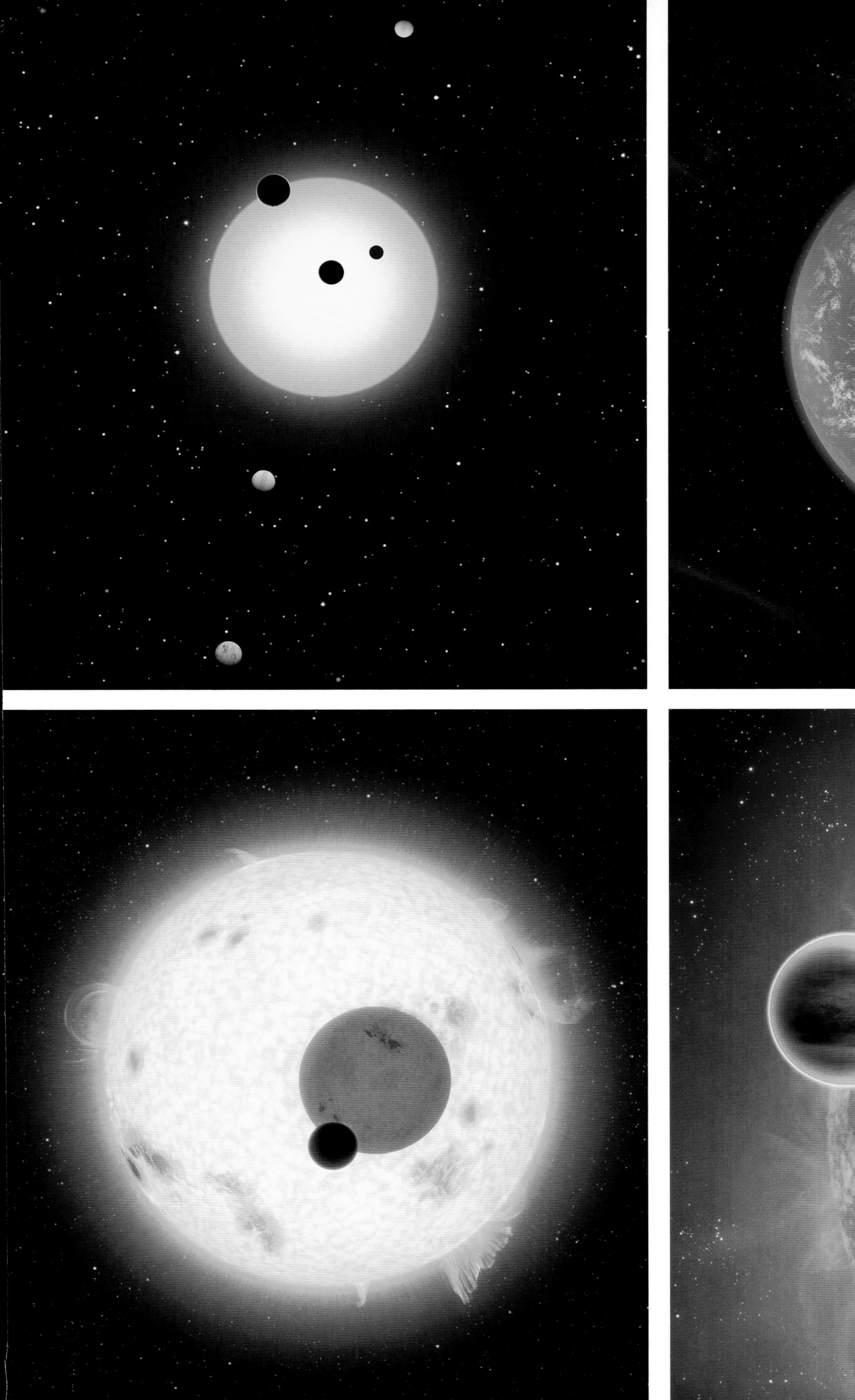

The reason for all this winking and wobbling is that most stars aren't drifting through the Milky Way in solitude. They're accompanied by planets, just as the sun is, and the winks and wobbles are the proof. If a planet passes precisely in front of its star from Earth's point of view, the star dims just a little. Even if things aren't lined up quite so perfectly, the planet's gravity will tug the star back and forth as it orbits, causing changes in starlight that scientists can see.

By measuring the winks and wobbles with exacting precision, astronomers have found literally thousands of alien worlds, known to astronomers as exoplanets, in an almost bewildering variety of sizes and orbits, since the first was discovered less than two decades ago. They haven't found one yet that truly resembles Earth, however—about the size of our home planet, orbiting at just the right distance from its star, with a temperature neither too hot nor too cold to harbor life.

But scientists believe they're on the verge of doing just that—and considering how short a time astronomers have been in the planet-hunting business, it will be an extraordinary achievement. "They discovered the very first planet orbiting a sunlike star in 1995, when I was in college," says Eric Ford, an astronomer at the University of Florida, "and now I get to help search for the first Earthlike planets. That's pretty cool."

It could also be transformative. Since the time of the ancient Greeks at least, philosophers have pondered and argued the question of whether humans are alone in the universe. While finding a second Earth won't answer that question immediately, it will take astrobiologists—scientists who work at the intersection of astronomy and biology—a big step closer by giving them somewhere to look.

The odds-on favorite for making such a discovery is the Kepler mission, a telescope whose sole function since its 2009 launch has been to stare ceaselessly at a field of some 156,000 stars located in the northern sky between the constellations Cygnus and Lyra, waiting for one or more of them to wink as a planet passes by. The idea was developed by Bill Borucki at the NASA Ames Research Center in California (and was rejected four times before NASA finally gave the project the go-ahead). Borucki reasoned if that you observe just one star hoping that a planet will pass it, you'll almost certainly fail, because planet and star have to line up with exquisite precision for the event to be visible. Look at many tens of thousands at once, however, and you're a lot more likely to catch a planet in the act.

But Kepler's job isn't just to find a second Earth. It's to see how common they are in the Milky Way, which is yet another reason to look at a huge number of stars at once. The Kepler has reeled in an impressive haul of planets along the way. At last count, it had snagged more than 2,000, including 204 Jupiter-size bodies; 988 Neptunes; 412 so-called "super-Earths" midway in size between Earth and Neptune; and 123 Earth-size planets. They aren't true twins of Earth because all of them are much too hot for life. But their existence tells astronomers that Earth twins must be out there in significant numbers.

BESIDE A MODEL OF THE KEPLER SPACECRAFT *at NASA's Ames Research Center, William Borucki, the mission's chief scientist, lectures on the space observatory orbiting the sun in search of distant Earthlike planets.*

Astronomers also know that the galaxy is teeming with all sorts of oddball solar systems that look nothing like our own. The first hints of this surprising development emerged with the discovery of the very first extrasolar planet back in the mid-1990s. The alien world known as 51 Pegasi b—found by Swiss astronomers Michel Mayor and Didier Queloz, who noticed that the parent star was wobbling in place—was "just weird," says Harvard planet hunter David Charbonneau. It was about half as massive as Jupiter, but much closer to its star than Mercury is to the sun. Its year was just four days long, something planetary theorists considered impossible for a giant planet.

Yet as new planets began to trickle in over the next few years, it became clear that so-called "hot Jupiters" weren't uncommon at all, and when the Kepler came online the strangeness just got stranger. In 2011, for example, the space probe found a planetary system around a star designated as Kepler 11 that contains no fewer than six planets, all of them significantly bigger than Earth, five of them crammed inside an orbit equivalent to Mercury's. "It's remarkable," astronomer Ford, a co-author of the discovery paper, told TIME. "We never expected to see something like this."

Astronomers also wouldn't have imagined Kepler 16b, a Saturn-size planet found late in 2011 that circles not one star, but a pair of stars that whirl around each other in a tightly bound orbit. Theorists doubted such worlds could exist in such a gravitationally unsettled environment—but again the universe didn't seem to care what theorists believed. "When we first saw it," said co-discoverer Joshua Carter of the Harvard-Smithsonian Center for Astrophysics, "I thought, 'Wow, this is just amazing.' It's hard not to get excited. This is too much fun."

The only problem with Kepler—if you consider it a problem—is that astronomers can't keep up with the flood of data streaming down from its powerful cameras. "The candidates it's finding," says David Latham, also of the Harvard-Smithsonian Center, "will keep the community busy for years." Of the 2,300 or so flagged so far by Kepler, only a couple of dozen have been studied in detail—and NASA recently gave the space probe the green light for four more years of operation.

The embarrassment of riches from Kepler coincides with many other planet searches going on using telescopes in space and on the ground—and those discoveries are also proving surprising and remarkable. The SuperWASP (for Wide Angle Search for Planets) survey, with scopes in South Africa and the Canary Islands, has found 26 planets of its own. Among them is WASP-12b, a world that's strangely rich in carbon. The planet is so big that it's probably made mostly of gas, but there could be smaller, solid planets nearby as well. "On a carbon-rich world," co-discoverer Nikku Madhusudhan, now at Yale, told TIME, "you could have big landforms made of pure diamond."

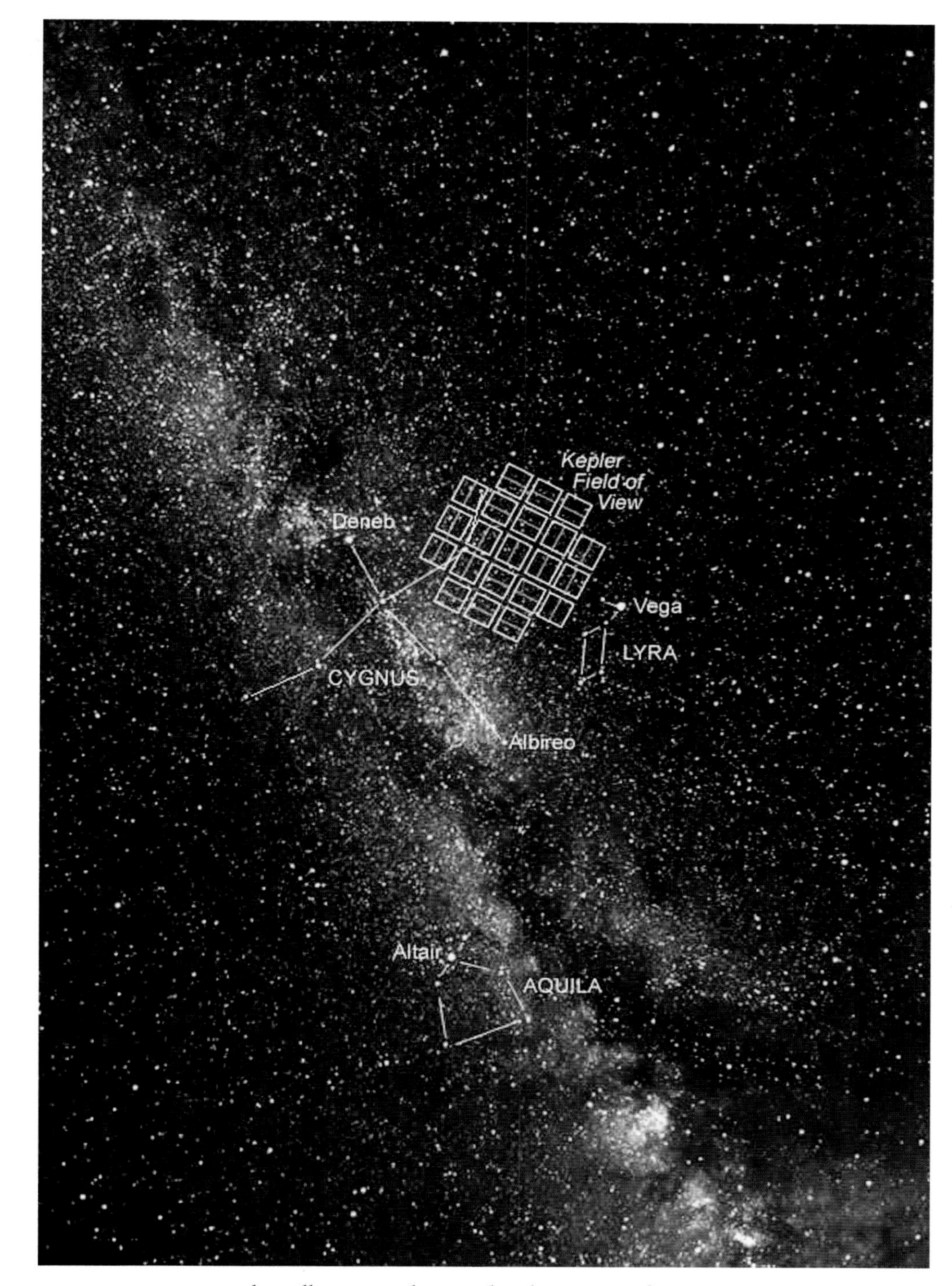

KEPLER TAKES AIM *at the Milky Way. Each rectangle indicates a specific target region covered by one of the 42 paired photographic CCD elements.*

Then there's the MEarth project, led by Harvard's Charbonneau, which focuses not on sunlike stars but on the much smaller, dimmer, redder and more numerous M-dwarf stars that make up 70% or more of the Milky Way. "I was taught in school that we orbit an average star," says Charbonneau, "but it's a lie. If the sun is a 100-watt bulb, most stars are like little Christmas lights." That's an advantage for planet hunters, though. If a planet passes in front of an M-dwarf, it blots out a bigger percentage of

EXOPLANET GJ 1214B *depicted (top) orbiting its small, reddish M-dwarf star, which is about 40 light-years from Earth. It is the first exoplanet whose atmosphere has been analyzed. Harvard's David Charbonneau (above) heads up the MEarth project, which led to the discovery of GJ 1214b.*

the tiny star's light, and since an M-dwarf is relatively puny, the planet's gravity can make it wobble more easily.

The strategy paid off in a big way with MEarth's discovery of a world known as GJ 1214b, a super-Earth discovered just a month before Kepler found its own first planet. Like the carbon planet found by SuperWASP, this one has a surprising composition: Less than three times the size of Earth, it is probably made of half rock, half water. "It's a top-of-the-top discovery in the quest for Earth-size planets," commented Berkeley's Geoff Marcy, the world's greatest planet hunter (if you don't count the Kepler, that is), when GJ 1214b was first announced.

But even a superlative discovery is destined to be topped in the fast-moving world of exoplanetology. In March 2012, a European team using a telescope in the high desert of Chile said it had found no fewer than nine super-Earths in a search of 108 nearby M-dwarf stars, including two in the stars' habitable zones. At that rate, there should be more than 3 billion such planets in the Milky Way—a number that would have seemed mind-blowing just a few years ago. Among his colleagues, though, says California Institute of Technology planet hunter John Johnson, "it was greeted almost with yawns, because it's gotten to the point where we're bored with super-Earths."

That is very probably an exaggeration. While astronomers are ultimately looking for a mirror Earth similar in size and composition to our own world (carbon planets and giant blobs of water need not apply), it's not clear that larger planets couldn't be life-sustaining as well. Says Dimitar Sasselov, director of the Harvard Origins of Life Initiative: "I don't see a dividing line [for planets friendly to life] anywhere between one Earth mass and five Earth masses and even 10 Earth masses." Johnson himself betrays a lack of boredom with planets in this size range: He's building a "micro-observatory" atop Mt. Palomar, in California, with four modest-size telescopes working in concert to find nearby habitable super-Earths. "We're just going to hammer away at the nearest, brightest stars," he says, "and basically shake the tree and see what falls out."

Still, while super-Earths might harbor life, the only place we know life

The artist's eye
No, this Earthlike world does not exist, but similar ones are surely out there

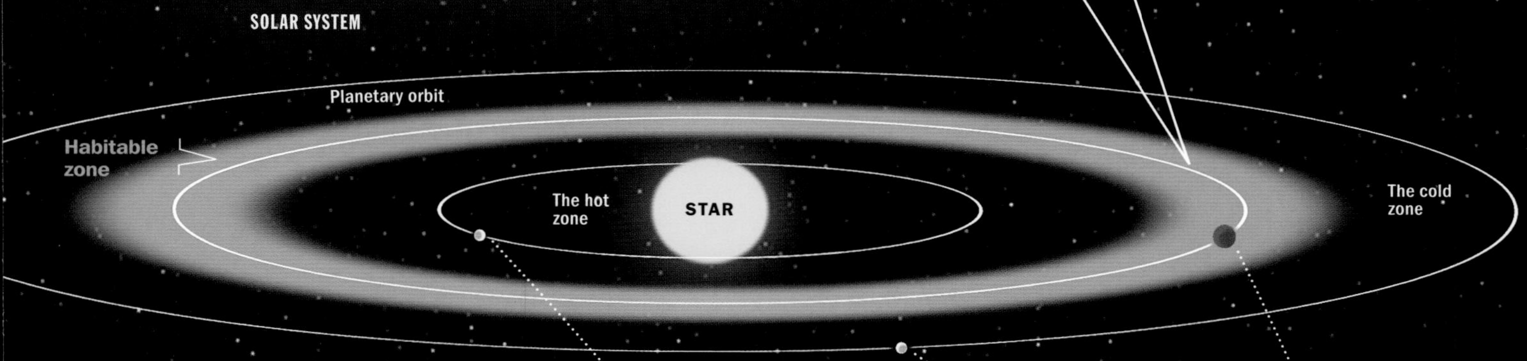

SOLAR SYSTEM

Planetary orbit

Habitable zone

The hot zone

STAR

The cold zone

Goldilocks Worlds: Where Things Are Just Right for Life

TOO CLOSE
The heat from a star can boil off water from planets that venture too close to it and warm their surfaces to deadly temperatures. Dry, airless Mercury and hothouse Venus illustrate the perils of proximity

TOO FAR
Space is a cold place, and you don't have to edge too far from your home star before water freezes solid. Atmosphere retains heat, and Mars might be a thriving world if it had held on to more of its air

JUST RIGHT
Earth exists in the habitable zone, where liquid water can be present in abundance. Life as we know it can't exist without water. Life as we haven't imagined it is, admittedly, more of a riddle

Three Ways to Spot Planets

1. WOBBLE
As a planet orbits, it gravitationally tugs its parent star this way and that. By measuring this motion, scientists can verify a planet's existence and infer its mass

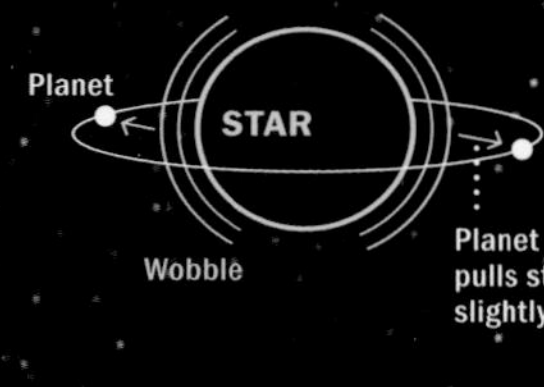

2. TRANSIT
Light from even the brightest star is slightly dimmed as an orbiting planet passes in front of it. The degree of dimming indicates the size of the planet

3. GRAVITATIONAL MICROLENSING
Gravity bends light, and a planet may hence distort the image of its star. This reveals the existence of a planet, but little more

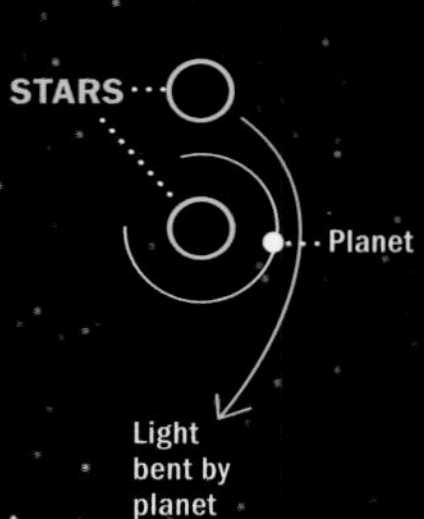

New Kids on the Block

For nearly all of human history, we've known only about the planets that circle our sun. In the past 20 years, scientists have detected hundreds of others orbiting distant stars. Most are giants compared with Earth, but improved detection methods are making it easier to find smaller ones

2,740

Candidate planets found so far by the Kepler Space Telescope as of January 2013. Here's the breakdown:

Neptune-size

1,290

Super-Earth size

816

Earth-size

351

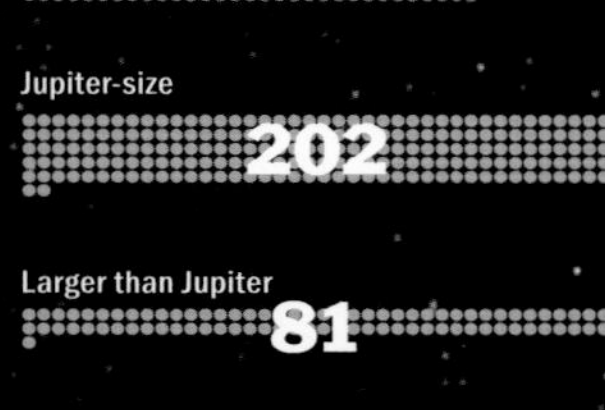

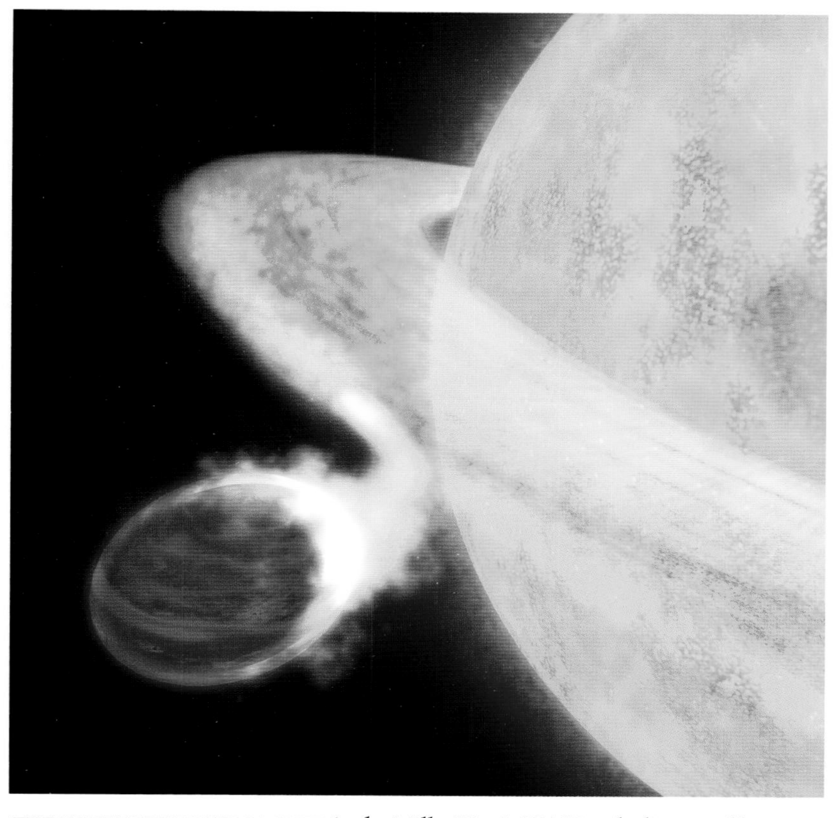

PLANET 55 CANCRI E *(rendered at left) orbits its sun so closely that it heats to more than 3,000°F. Life-sustaining planets would be thousands of degrees cooler and difficult to detect, but Caltech astronomer John Johnson (above) believes they could exist.*

THE HOTTEST KNOWN PLANET *in the Milky Way is WASP-12b, discovered by SuperWASP (for Wide Angle Search for Planets). The carbon-rich exoplanet is so close to its yellow dwarf star that it superheats to 2,800°F and bulges due to tidal forces.*

exists is on Earth itself, so a true twin of Earth remains the smartest kind of planet to seek. Astronomers don't have any of these in hand—yet—but with all the searches now going on, they likely soon will. Once that happens, scientists will need to observe these Earthlike worlds directly: You can prove a planet is there via winks and wobbles, but to find life, you've got to study its atmosphere for molecules such as ozone, which are telltale signs that life might be present.

That's a very tough task for existing telescopes. Astronomers have done some studies along these lines by watching as a planet transits in front of its star and noting how the star's light is altered as it passes through the planet's atmosphere. But these have all involved planets close-in to their own sun, too hot for life. To look for life on a habitable planet, you'd have to observe it directly. In May 2012, the infrared-sensitive Spitzer Space Telescope took a step in that direction by snagging a direct image of a super-Earth called 55 Cancri e. Here too, however, the planet was close to its sun: It was visible because the star heats it to more than 3,000°F. A habitable planet, one that's thousands of degrees cooler, would be invisible to any conventional telescope.

That's why planet hunters are frustrated that NASA has put the flagship Terrestrial Planet Finder (TPF) mission on the back burner for now. As originally conceived in the late 1990s, TPF would have involved four huge space telescopes flying in tight formation out in the neighborhood of Jupiter. Since then, astronomers have come up with a far less ambitious design—a single big telescope armed with technology to blot out the light from a bright star so the dimmer planets around it can shine through. But even that slimmed-down version is too expensive for a cash-strapped NASA to handle at the moment.

There's a chance, however, that the James Webb Space Telescope, slated for launch by 2018, can take on some planet-imaging duties until TPF is finally authorized. Given the astonishing progress planet hunters have made over the past few years in finding worlds closer and closer to the size and temperature of Earth, there will almost certainly be plenty of candidates for it to look at.

It's still an open question whether any of them will betray the presence of life—but the answer to that question is nearly within reach. "Over the past 15 years," says Johnson, "this idea has moved from the realm of science fiction to credible discussion. There are a lot of people left over from the old days who will still scoff at the notion. But we now know that the galaxy is teeming with Earth-size planets. I mean, we know this."

It could be, of course, that not one of these billions of planets gave rise to life the way Earth did. But very few astronomers would call that a safe bet.

CHASING THEIR TAILS

TIME WAS, IF PEOPLE THOUGHT ABOUT COMETS AT all, they thought Halley's. If you missed its once-every-76-year return—well, too bad. But comets, we now know, are everywhere. The Kuiper belt and the more distant Oort cloud, which surround the solar system, are home to massive swarms of comets and other icy, rocky bodies. Some break free and dive-bomb the sun, sweeping in, lighting up and then disappearing into the deep again. That's how Halley's comet came to be, and that's the provenance of the less spectacular ones too. Of course, less spectacular doesn't mean less interesting—at least not to astronomers—and comets have become a go-to destination for space probes. In 2005, NASA's Deep Impact spacecraft fired a projectile into the nucleus of comet Tempel 1, kicking up a spray of ice and dust for analysis and gouging a pit that offered a glimpse at the comet's innards. The Stardust probe flew by comet Wild 2 in 2004 and collected a sample of its corona. Multiple comets may have collided with Earth a long, long time ago, and though we weren't here to experience them, we might not be here at all if the collisions hadn't happened. Studies of the chemical signature of the ice in comets reveal that it matches up nicely with the water in our oceans. We're an exceedingly wet world now, but we may have been an exceedingly dry one once—until a few kind comets delivered all the water we'd ever need.

COMET MCNAUGHT, *the whoosh at far left (photographed in January 2007 near the Chiro Observatory in western Australia) was one of the brightest comets of the past 50 years—so bright it could be seen with the naked eye.*

NEW EYES ON THE UNIVERSE

BY DAVID BJERKLIE

Telescopes of unprecedented power and precision are enabling astronomers to peer ever deeper into the farthest reaches of space and time

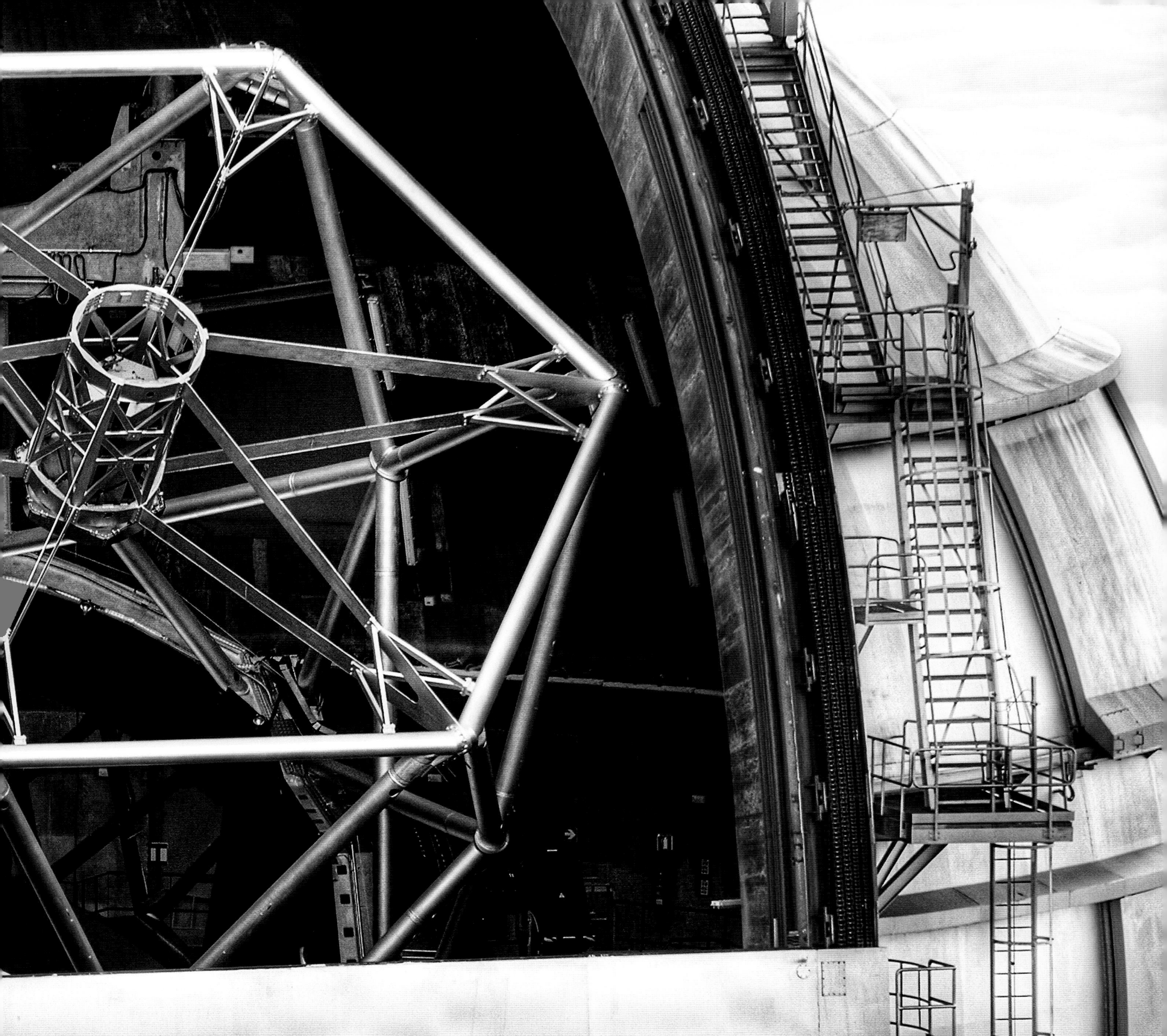

THE GREAT CANARY TELESCOPE, *which sits atop a mountain on La Palma in the Canary Islands, 150 miles off the Moroccan coast of Africa, is the world's largest single-mirror optical telescope.*

LARGE BINOCULAR TELESCOPE, MOUNT GRAHAM, ARIZONA

Space telescopes get the press, but mountaintop observatories still play a role. The Large Binocular Telescope (LBT) at left (exterior and interior, from top) has a pair of mirrors, each 8.4 m (27.5 ft.) in diameter, that give it unparalleled light-collecting power. The latest in sensors, software and adaptive optics correct for atmospheric blurring. The result is images whose detail and sharpness exceed current space-telescope capabilities. The LBT studies the formation, structure and rotation of galaxies and the massive black holes that power quasars.

VERY LARGE TELESCOPE, CERRO PARANAL, CHILE

The Paranal Observatory is situated in the isolated Atacama desert of northern Chile, which has a near-perfect combination of dry air and dark night skies. The centerpiece of the observatory (above) is the Very Large Telescope (VLT), which consists of four large mirrors, each 8.2 m (27 ft.) in diameter, plus four smaller, movable mirrors. The VLT has produced the first image of an extrasolar planet, tracked the movements of stars near the supermassive black hole at the center of the Milky Way, and measured the age—13.2 billion years—of the oldest star in our galaxy. To minimize light pollution, the observatory residence for staff and guests, including the swimming pool (top), is underground.

LARGE SYNOPTIC SURVEY TELESCOPE, CERRO PACHÓN, CHILE

The vastness of the cosmos poses a stark challenge: Powerful telescopes can zoom in on details, but how can they cover more than a tiny slice of the entire sky? The Large Synoptic Survey Telescope (LSST) will use innovative technologies, including the world's largest digital camera, to rapidly scan a field of unprecedented width. This will allow it to trace billions of remote galaxies and to track objects that change or move, from exploding supernovas to near-Earth asteroids. A technician (left) looks for flaws in the LSST's mirror; each segment is polished to within one-millionth of an inch.

ATACAMA LARGE MILLIMETER ARRAY, CHAJNANTOR PLATEAU, CHILE

While visible light reveals the stars in the galaxies, the Atacama Large Millimeter Array (ALMA) can "see" the massive clouds of dense, cold gas from which new stars and planetary systems form. Currently operating but not yet at full capacity, ALMA (rendered below) will consist of 66 high-precision, movable radio antennas; the ability to reposition them across northern Chile's high-altitude desert plateau will give the array maximum flexibility. One of ALMA's most striking images so far is a pair of colliding cosmic giants called the Antennae galaxies.

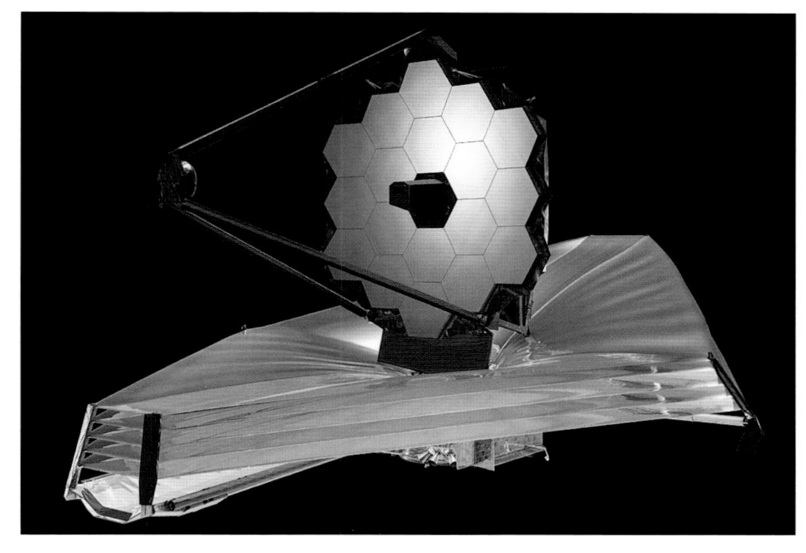

JAMES WEBB SPACE TELESCOPE
Expected to launch in 2018, the James Webb Space Telescope (above) should surpass its predecessors Hubble and Spitzer. Like the Spitzer, the Webb is an infrared telescope. Its missions will be to search for the first galaxies formed after the Big Bang and to study the evolution of stars and planetary systems. The Webb's mirror (top) will consist of 18 primary mirrors that will unfold, as in the rendering above, in orbit.

SOUTH AFRICAN LARGE TELESCOPE, SUTHERLAND
Situated in a wildlife reserve, the South African Large Telescope (SALT) can detect objects a billion times too faint to be seen with the unaided eye (imagine how a candle flame on the moon would appear from Earth). SALT's ability to take snapshots in rapid succession allows astronomers to study the complex dynamics of binary star systems. In the long-exposure photo of the telescope at left, the Earth's rotation makes the stars appear to move in circles.

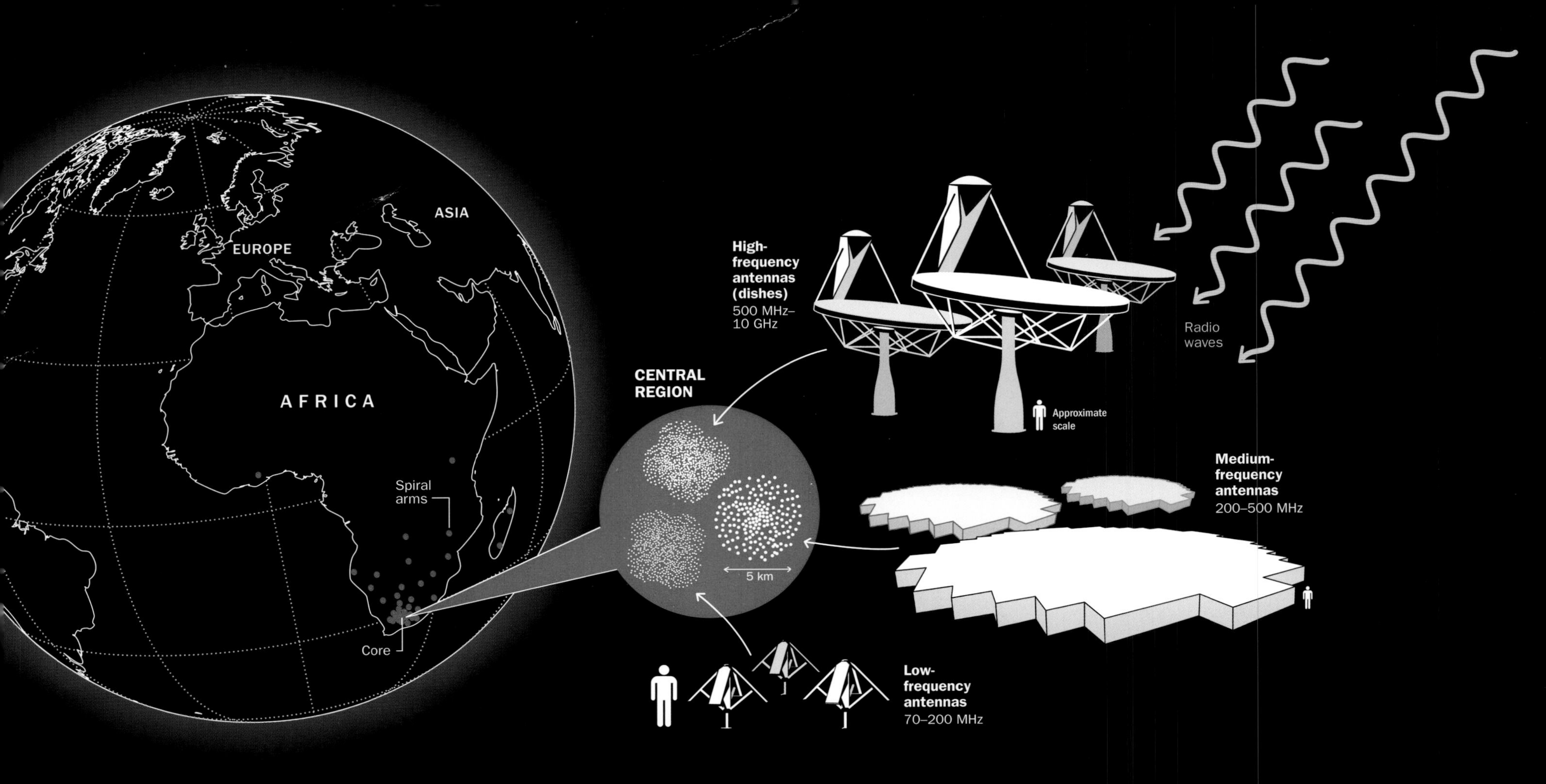

Peerless

The largest and most sensitive telescope ever built, the Square Kilometer Array will peer deeper into the universe than ever before, shedding light on some of its greatest mysteries

1 **Three types of receptors**—familiar dishes and two less familiar shapes—all with different frequency sensitivities, will be **densely distributed** in the central region of the array

2 **Additional groups** of the receptors will be placed along **spiral arms** spanning as far as 3,000 km from the central core. The larger collecting area boosts the telescope's sensitivity

3 All antennas will connect to a **supercomputer that combines the signals,** creating an unprecedented view of the sky. The array will generate enough raw data to fill 15 million 64-GB iPods every day

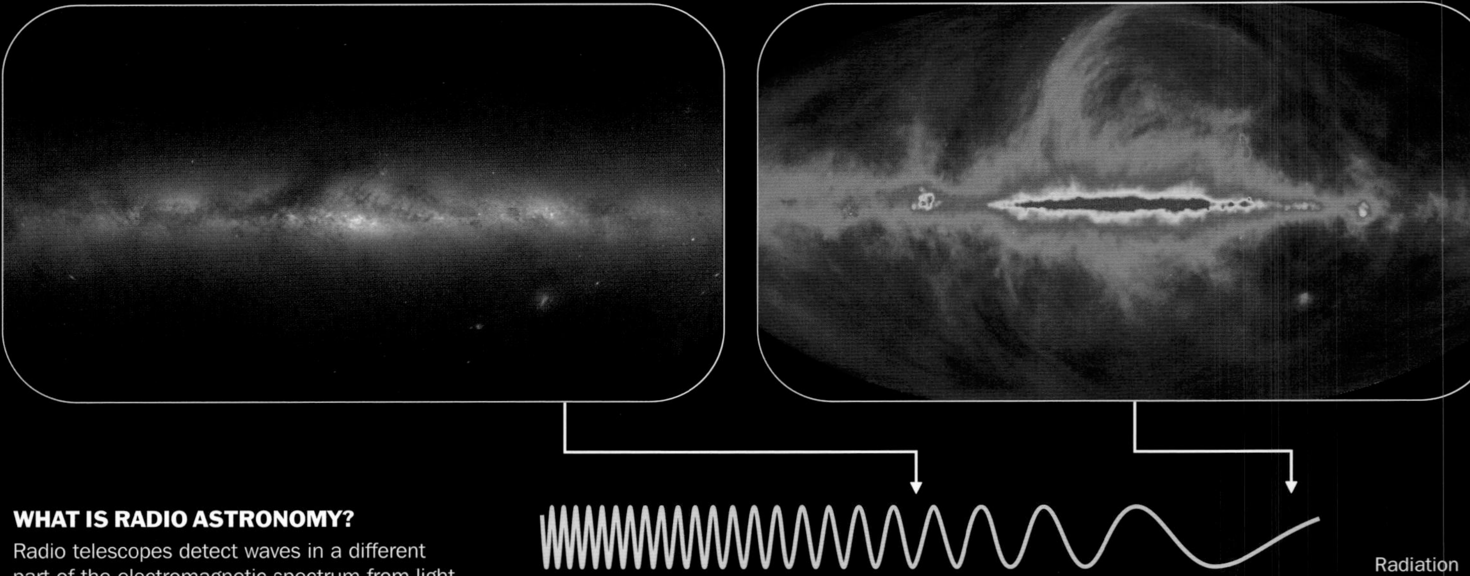

WHAT IS RADIO ASTRONOMY?

Radio telescopes detect waves in a different part of the electromagnetic spectrum from light waves, which optical telescopes can detect

Big Questions the SKA Could Help Answer

IS ANYONE OUT THERE?
The telescope will enable scientists to detect very weak radio signals from far away, perhaps indicating intelligent life.

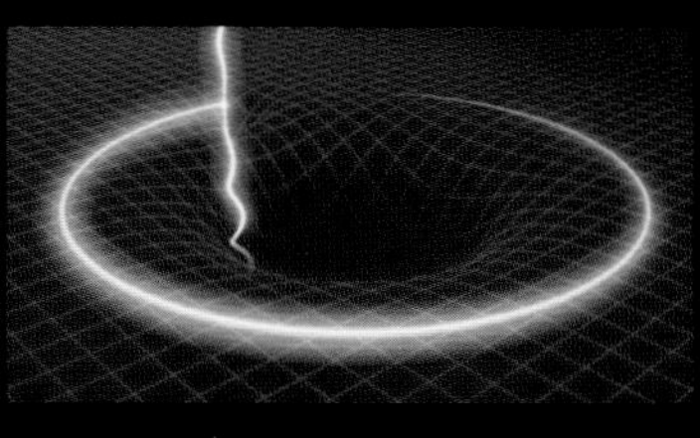

HOW DID BLACK HOLES FORM?
The SKA's strength will illuminate a relatively narrow gap in cosmological history when the first stars and black holes formed.

WHAT IS DARK ENERGY?
The SKA will detect a billion galaxies, allowing scientists to study the puzzling force that drives the universe's expansion.

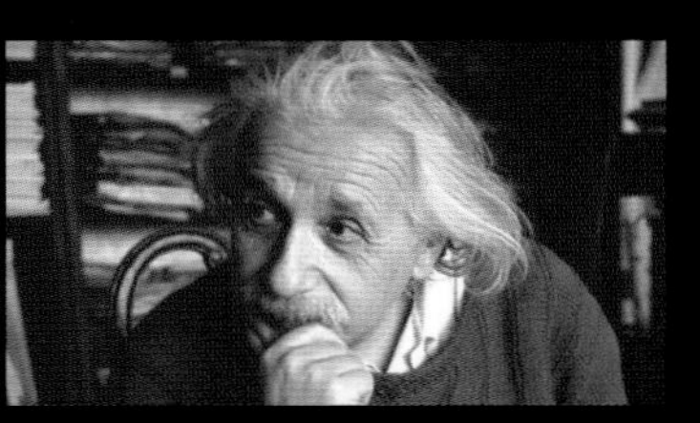

WAS EINSTEIN RIGHT?
Astronomers will be able to study space and time in extremely curved regions of space, testing the Theory of General Relativity.

SOURCE: SKA ORGANIZATION

AFRICA'S EYE ON THE SKY

BY ALEX PERRY/JOHANNESBURG

A new radio telescope could provide radically new insights into the Origins of Everything. But it's the place the instrument may be built that's creating the most excitement

FOR SOMEONE WHOSE JOB TITLE COULD READ "MAN MOST LIKELY TO BLOW YOUR MIND," Bernie Fanaroff looks pretty conventional. Short, affable and 65, Fanaroff wears a V-neck and gray slacks and offers coffee and sandwiches when we meet in his windowless office on a busy thoroughfare in central Johannesburg. Then he opens his mouth, and out tumbles all manner of cosmic craziness. Consider the fact, says Fanaroff, that we have no idea what 96% of the universe is made of. Cosmologists have known for some time that only 4% of the universe is stuff like dust, gas and basic elements. Dark matter, says Fanaroff, accounts for 23% to 30%; dark energy makes up the rest. (*Dark,* Fanaroff explains, is the scientific term for "Nobody knows what it is.")

Fanaroff is consumed by such fundamental mysteries, to say nothing of the larger riddles surrounding the Big Bang. It's to investigate all this—to answer questions about the Origin of Everything—that he is working on the Titanic telescope, an astronomical instrument 50 to 100 times more sensitive than anything yet created. It will see so far across space—scanning a billion galaxies, up from the currently observed million, with an ability to pick up the energetic equivalent of an airport radar on a planet 50 light-years away—that it might even find evidence of aliens (if they happen to be in range and using radio-emitting devices). It will see so far back in history that it will transport astronomers closer to the Big Bang than any previous telescope. It will help answer questions such as whether Einstein's Theory of General Relativity holds, what cosmic magnets look like and whether black holes are hairy. (Honestly. It's in a brochure Fanaroff hands me.) It will also, it is hoped, massively accelerate global computing power, as the volume of data it will generate will be greater than the amount so far created in all human history and more than the entire Internet can handle. And if all that sounds a little spacey, get this: Fanaroff's hoping to build the thing in Africa. "It's a massive leap into the unknown," he announces, somewhat superfluously.

Fanaroff's real title is project director. The proper name for the Titanic telescope is the Square Kilometer Array (SKA) radio telescope—so called because it will be made up of thousands of radio-

THE SQUARE KILOMETER ARRAY (SKA) DISHES *rendered here are part of a multinational scientific initiative designed to answer questions about the origins of the universe. Located in Australia, New Zealand and Africa—and headquartered in England at the Jodrell Bank Observatory—the sophisticated radio array will operate over multiple frequencies, making it 50 times more sensitive than other radio telescopes and capable of surveying the sky 10,000 times faster than previous instruments.*

frequency receivers with a collective area equal to 1 sq. km. Some of the receivers are 12-meter-diameter dish antennas, some are smaller, and some are flatter, fish-eye constructs. Each targets a different frequency range to deepen and expand astronomy's view of the sky. That kind of stargazing power requires a lot of real estate here on Earth. The SKA plan calls for 20% of the antennas to be clustered within a 1-km radius, 50% within 5 km and some as far as 3,000 km away—putting them in eight other African countries as far away as Ghana and Kenya and across the Indian Ocean in New Zealand and Australia.

Scattering so many dishes and collating their images into one composite will produce the same effect as building a single dish half the size of Earth. Why the obsession with size? Because with telescopes, bigger is always better—improving resolution and helping triangulate more precisely on cosmic targets—and this one is meant to be the best ever built.

Astronomers can already see back 13.5 billion years, about 400,000 years after the Big Bang, by looking at light very far away and therefore very long ago. But they get no further. Why? Because the Big Bang was extremely hot, and only after around 400,000 years did it cool enough to allow protons and neutrons to form into atoms—most importantly hydrogen, the first element. Before that 400,000-year point, what existed was a very hot, dense primordial fog. And we can't penetrate that with optical devices like the Hubble Space Telescope or the European installation in Chile's Atacama Desert called the Very Large Telescope. (Remember that obsession with size?) But what a radio telescope—particularly a very big, very sensitive one—can see into is a somewhat more recent foggy period known as the Dark Age, and mysteries aplenty lurk there. Hence the supersize SKA.

Something as big as half the world needs much of the world to fund it. The SKA is a collaboration of more than a dozen countries that will share its estimated building costs of $2 billion. When bidding to host the SKA began in January 2003, the U.S., China, Australia, Argentina and South Africa all expressed interest. But the best place for a radio telescope is way out in the middle of nowhere, with no interfering radio signals, particularly no mobile phones. That ruled out the U.S., China and Argentina and left the great empty deserts in western Australia and northern South Africa. In May 2012, the funding nations recommended splitting the SKA primarily between those two. Said Jim Cordes, a professor of astronomy at Cornell University: "Most people see the tremendous work going on in both countries, and we would want to keep everyone involved." Benefits of hosting include hundreds of millions of dollars of investment in power, communications and data processing, not to mention the prestige of being home to a project that Fanaroff expects will produce "several Nobels." Everyone involved, from developers to funders to host countries, could also cash in on business spinoffs that could come from the work. (Wi-fi and digital cameras are among products originally developed by astronomers.)

For South Africa in particular, hosting the SKA is a game changer. When Fanaroff's team submitted a bid, the world of science, he says, was "a bit taken aback." He sees his mission as partly to overturn such "Afro-pessimism," and he is succeeding: In the years since, surprise has become admiration. His program of training engineers and physicists—with special attention to women and blacks—has produced a team of astronomers widely acknowledged to be equal to any in the world. Brian Boyle, head of the Australian bid, sees the decision to place the world's most powerful telescope in the southern hemisphere as "a democratization of science."

The SKA will not be complete until after 2020. But even before the May 2012 announcement, construction was underway on smaller telescopes that will eventually be incorporated into the full SKA. By early 2012, the first seven South African dishes were already live and transmitting back to a monitoring station in Cape Town. When I visited the station, telescope manager Willem Esterhuyse showed me a printout of the first readings. Cordes from Cornell had told me to prepare to be surprised. "The origin of life is one thing," he had said. "This is the Origin of Everything." Something very far away, perhaps even the Big Bang, was speaking to me on the paper I held in my hand. But I didn't see black holes, and I didn't see the Dawn of Everything. Instead, what I saw was a series of rainbows of varying width, similar to the test printout from a new color copier. "It makes you think, doesn't it?" asked a grinning Esterhuyse.

Like I said: spacey.

CAMBRIDGE UNIVERSITY'S BERNIE FANAROFF *is SKA's project director. He calls it an "iconic project—the biggest research infrastructure in the world."*

E.T., ARE YOU CALLING US?

BY MICHAEL D. LEMONICK

The discovery of Earth-size planets has made scientists more confident than ever that we are not alone. But finding the money to pay for the search is nearly as challenging as the hunt for aliens

From its very beginnings more than six decades ago, SETI, the Search for Extraterrestrial Intelligence, has been a science fueled by hopes and dreams but not a particle of solid evidence. A young astronomer named Frank Drake launched the field with what he called Project Ozma in 1960 by pointing his radio telescope at the star Tau Ceti in an effort to detect an alien broadcast signal. But neither he nor anyone else has ever verified a single peep from an extraterrestrial civilization.

No one even knew back in the 1960s whether

THE ALLEN TELESCOPE ARRAY (ATA) *at the Radio Astronomy Laboratory at UC Berkeley utilizes large numbers of small-diameter dishes (LNSD) instead of a more costly single dish. Equivalent in sensitivity to a single 100-meter-diameter dish, ATA is employed in the SETI program to monitor direct signals and other signs of transmissions from civilizations on other worlds.*

JILL TARTER *(in the Arecibo Observatory dish control room in Puerto Rico) recently retired as director of the Center for SETI Research in order to seek private funding for the cash-strapped project.*

planets existed beyond our own solar system—or, if they did, whether they had given rise even to primitive life, let alone to beings who could beam radio waves trillions of miles across the vast emptiness of interstellar space. And even if they had, who was to say for sure that those beings used radio for communication? The odds seemed so long that Drake would advise young scientists interested in SETI to do other research as well, so they'd have a better chance of actually discovering something. Otherwise, he felt, it could be too disappointing.

But plenty has changed since those early days. We know there are planets orbiting other stars. Nearly 3,000 so-called "exoplanets" have been found since the first one was discovered in 1995, and to everyone's astonishment, many of them revolve around stars much smaller and redder than our sun—places nobody had considered before. It's clear that this is merely the tip of the cosmic iceberg; it's also clear, say the experts, that the discovery of a true Earthlike world, is probably not far off. "This is revolutionary," says Jill Tarter (the real-life inspiration for Jodie Foster's character in the film *Contact*), who recently retired as director of the Center for SETI Research in Mountain View, Calif. "Now we can point our telescopes at stars where we know planets exist."

Not only that: The electronics that SETI searchers have at their disposal have seen the same transformation that packs more data-processing power into your smartphone than there was in a room-size 1960s-era computer. "We've got much more powerful receivers," says Frank Drake, who is now 82 and technically retired but hasn't slowed down a bit. "It's routine now to search through 100 million radio channels at once using cheap off-the shelf components."

In 1998, moreover, searchers began looking not just for radio signals but for the bright flashes of alien lasers in what's now known as Optical SETI, or OSETI. "On Earth," says Berkeley's Geoff Marcy, who has found more exoplanets than anyone else, "we've now got lasers that can emit more than a petawatt [a quadrillion watts] of power for up to a nanosecond." Any self-respecting alien civilization would presumably have them too, and says Marcy: "They're so bright they can outshine a star."

With this expanded search strategy, more powerful detectors, and a better fix on which stars to target, you might assume that we've finally entered the golden age of SETI—especially since the 2007 debut of the Allen Telescope Array (ATA) in Northern California, the first radio installation built expressly for SETI. Before the Allen, searchers had to compete for time on existing radio telescopes.

But for all the many things going right with SETI, one shadow has hung over the program for much of its existence and is looking more ominous than ever: a terrible shortfall in funding. For a while back in the 1980s, NASA was putting money into SETI—naturally enough. But in 1993, Nevada's Sen. Richard Bryan, declaring that "not a single Martian has been found," introduced an amendment cutting off all NASA support. "He was trying to make a name for himself as a budget hawk," says Drake ruefully.

GEOFF MARCY, *the director of Berkeley's Center for Integrative Planetary Science and co-investigator for NASA's Kepler mission, has discovered more exoplanets than anyone in history.*

Since then, the search has depended on private money. "We were getting donations from people like Bill Hewlett and Dave Packard, the founders of Hewlett-Packard," says Drake, "and Gordon Moore, the co-founder of Intel, and Paul Allen, the co-founder of Microsoft. Paul Allen gave $30 million for the Allen Telescope Array." But Hewlett and Packard passed away, and when the economy tanked in 2008, says Drake, fortunes worth $30 billion were suddenly worth only $20 billion. Overall, private funding has plummeted dramatically.

Berkeley, meanwhile, which had agreed to fund the telescope's operations if the SETI Institute raised the money for construction, went through its own financial crises and pulled back. The ATA had to go offline in 2011. "If you think of SETI as not just research but exploration," SETI Institute senior astronomer Seth Shostak said at the time, "this is like sending Captain Cook to the South Pacific but not giving him any food or supplies." Thanks to a fundraising campaign, the array's 42 radio dishes came back to life earlier this year, and the Air Force has bought some of the telescope's time to keep tabs on orbiting space junk.

But money is still very tight, and Tarter stepped down from her position in part because she wanted to devote her full time to bringing in desperately needed cash. "It's a dramatic juxtaposition," says Marcy, "between a compelling science quest and an utter lack of resources." That's not only true for the SETI Institute, but also for the handful of other small-scale searches going on in the U.S. and in a half-dozen other countries, including Italy, Argentina and Japan.

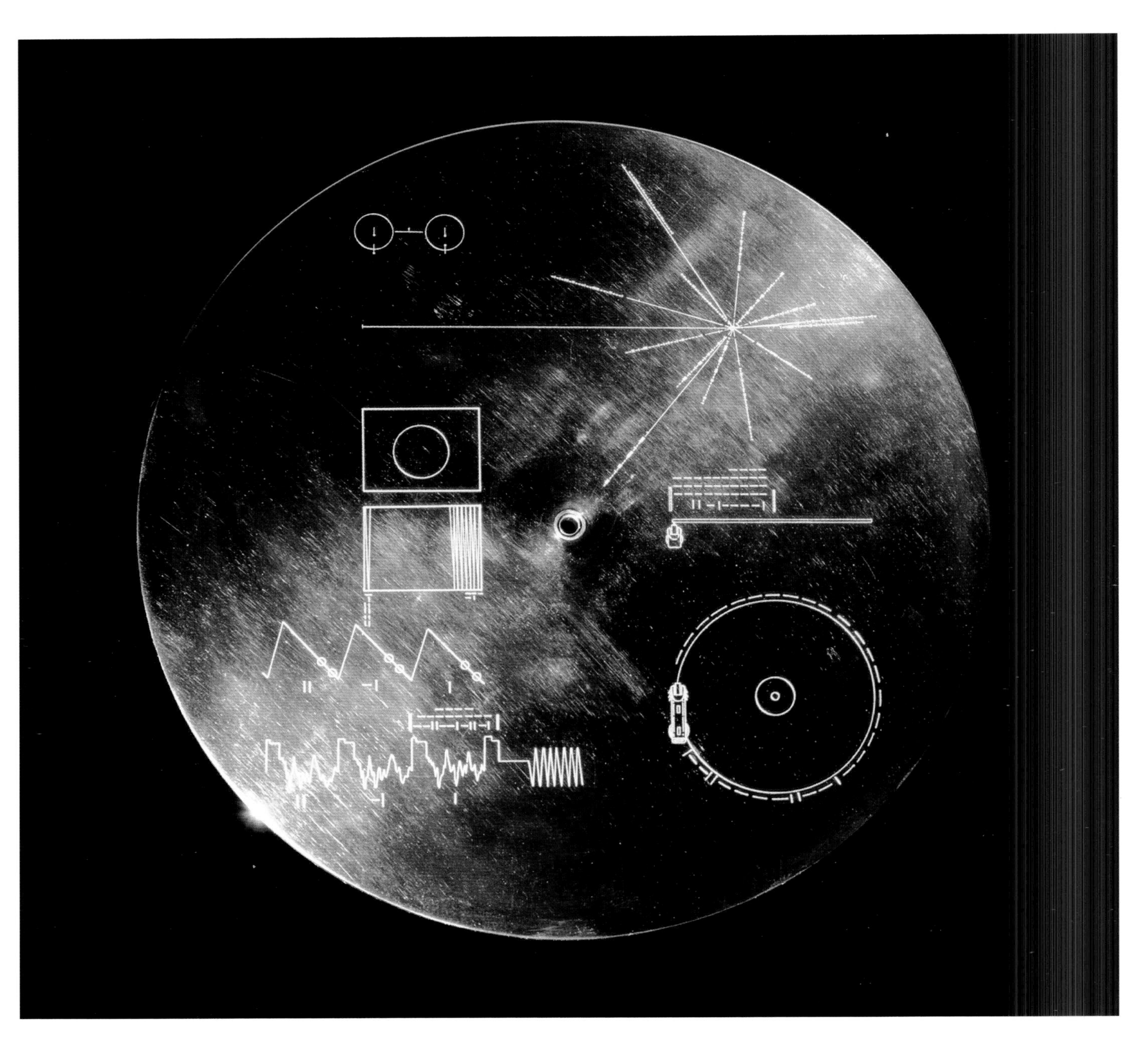

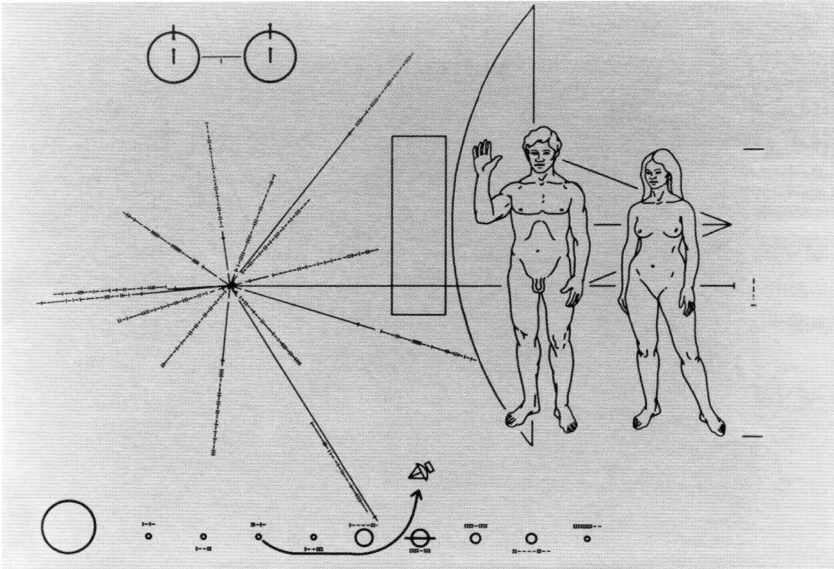

EARTH'S COSMIC INVITATIONS
Space probes Pioneer 10 and 11, which launched in 1972 and 1973, respectively, were outfitted with gold-anodized aluminum plaques featuring anatomically correct male and female humans (left) along with other symbols and information useful to locate Earth. The gold-and-aluminum-cased "Sounds of Earth Records" (above) took the concept further. Mounted on twin probes Voyager 1 and 2 launched in 1977 were gold-and-copper-plated records containing nature sounds, a variety of cultural music, and greetings in 60 languages. As the probes travel farther from our solar system into interstellar space, might there be an alien meet and greet within the next few years?

ASTRONOMERS
Shelley Wright, Frank Drake and Remington Stone flank the Lick Observatory's 1 m (40-in.) reflector telescope, which has been modified to detect nanosecond laser pulses from beyond our solar system.

VOLUNTEERS CAN *now join the hunt for extraterrestrial intelligence. Visit setiathome.berkeley.edu to download software that delivers data from SETI's radio telescopes, analyzes it and then brings the findings back to SETI. The program is working when you aren't—during screensaver sleep mode.*

Perhaps the private sector has just gotten tired of SETI, Marcy speculates, or maybe it's just too difficult to overcome the "giggle factor" that has dogged the search right from the beginning—understandable enough, given the public's exposure to UFO nuts and outlandish sci-fi movies. Or maybe, he says, the public needs an inspiring proponent. Says Frank Drake: "Anytime I've done anything in the media, I've hoped it would stimulate interest. But [the late] Carl Sagan," an early and enthusiastic SETI booster, "was a lot more eloquent."

Given SETI's money woes, you might expect its proponents to think about giving up. But that's clearly not happening. Marcy, for example, is eager to ratchet down his planet-hunting research so he can get more deeply into SETI himself. "Thanks largely to the Kepler mission," he says, "we now know that there must be Earth-size planets in their stars' habitable zones." Once you have a rocky planet with plenty of water and lukewarm temperatures, he points out, "the biochemistry of life is a foregone conclusion. Yes, intelligent life is another matter, but life is more or less a done deal."

That being the case, and assuming primitive life gives rise to intelligence at least some of the time, Marcy has already begun his own modest search, looking for evidence of extraterrestrial lasers. He's not simply watching for flashes of light, however. He's using a spectrometer, which smears starlight into the rainbow of colors its light is made of. A laser, by contrast, emits just one very specific color of light (the first ones were red, but lasers now come in green, violet, yellow and blue—hence the term Blu-ray). If an alien civilization were flashing lasers at us, the color of the lasers would show up far brighter than the rest of the star's colors. "The spectroscopy we do of stars during our planet searches," says Marcy, "is amenable to this technique. My plan is to go back to all of our observations and look for these colored SETI dots."

He's got another idea as well. "This is clearly harebrained," he says. "But think of what happens when you point at the nucleus of another galaxy. You're now seeing 50 billion stars at once—it's the metropolitan downtown area." You're also looking millions of light-years away, too far for a conventional laser to be visible. But in a few rare cases, extraterrestrials might be so advanced that they can tap into the energy of the giant black hole that lurks in every galaxy's core. "If so," says Marcy, "they could power truly amazing lasers, visible halfway across the universe. We've already taken spectra of 35 or 40 galaxies, looking for this effect, and I'm serious enough that I've applied for more telescope time."

Others want to look for laser-flashing aliens in a different way. As an undergraduate at the University of California, Santa Cruz, a little over a decade ago, Shelley Wright got her initiation into SETI working on an optical search project with Frank Drake. Now on the faculty of the University of Toronto, Wright—perhaps mindful of Drake's warning to young scientists—works on other problems in astrophysics such as the evolution of galaxies, but she has never lost her passion for SETI. Thanks to some funding she got with Geoff Marcy's help, Wright is designing a detector that can pick up bursts of infrared light.

One reason for choosing infrared is that many stars shine more dimly in the infrared spectrum than they do in visible light. A conventional visible-light laser might be 1,000 times brighter than a star, bright enough, observes Drake, that—if our eyes could detect pulses as short as a nanosecond—"you could literally sit in the backyard at night and see extraterrestrial beacons, if they were there." But an infrared laser, according to Wright, could outshine a star by a factor of 10,000: "We won't be looking for a needle in a haystack," she says. "We'll be looking for a fire." An even bigger advantage, she explains, is that infrared light is much better than ordinary light at shining through clouds of interstellar dust. "We hope to get our infrared instrument built and on the sky in the next year or two."

"We" still includes Drake, who stopped following his own advice about working on anything other than SETI many years ago. He's collaborating with Wright on her infrared detector, for example, and he's also thinking about new projects that he and his colleagues might undertake if they could get their hands on more funding. Military radar, for example, deliberately hops from one radio channel to another in a random sequence, so it can't be jammed. If aliens were doing that with their signals, says Drake, "we couldn't detect that. The problem could be solved, but it takes more hardware."

Drake also wants to look backward. "There have been hundreds of searches since 1960," he says, "and many have reported candidate signals, but when people went back to look, there was nothing there." Drake and others suspect that at least some signals might have been real, but that by the time of the follow-up observations, which often came days or weeks later, the aliens had aimed their beacons elsewhere.

The most celebrated of these missed opportunities: the so-called "Wow! signal." In 1977, an antenna in Ohio picked up a radio blip lasting more than a minute that seemed as though it could be from alien broadcasters. Unfortunately, scientists didn't notice the signal in the telescope's computer output until several days later. An Ohio State astronomer wrote "Wow!" on the printout—but the signal never repeated. "We need to have equipment set up in such a way that we can do immediate follow-ups," says Drake. The SETI Institute recently created an online program called SETILive, which lets ordinary citizens monitor the Allen Telescope Array's computers for signals that merit follow-up. But that's not nearly enough. "We need to increase funding," he says, "so that we've got personnel ready to leap on a promising signal."

All of these ambitious searches, like Drake's very first during the last days of the Eisenhower administration, make one key assumption: There's someone out there. But in fact there's another assumption that often goes unmentioned: In order for a radio or a laser signal to be detectable across tens or hundreds of light-years, it has to be aimed more or less right at Earth. "It would have to be intentional," he says. And because communication between the stars is expensive in terms of both hardware and energy—and also because it wouldn't be likely to benefit the aliens—it would have to be altruistic. "This raises the question," says Drake, "of whether altruism is widespread, or whether it's a rare thing." Maybe humans are freaks for being altruistic even some of the time. "My own opinion," he says, "is that it's probably widespread, because it would be selected for by evolution."

IN 1977 AN ANTENNA IN OHIO PICKED UP THE "WOW" SIGNAL, A RADIO BLIP OF MORE THAN A MINUTE THAT IT SEEMED COULD BE FROM ALIEN BROADCASTERS

Whether humans are curious enough to spend more money to find out, however, is perhaps an even greater unknown. Says Geoff Marcy: "SETI has to become a major effort in the 21st century. It has to rise into the center of the radar screens of NASA, the National Science Foundation, the European Space Agency."

The truth, in short, is out there. But it'll take a bit more cash to learn what it is.

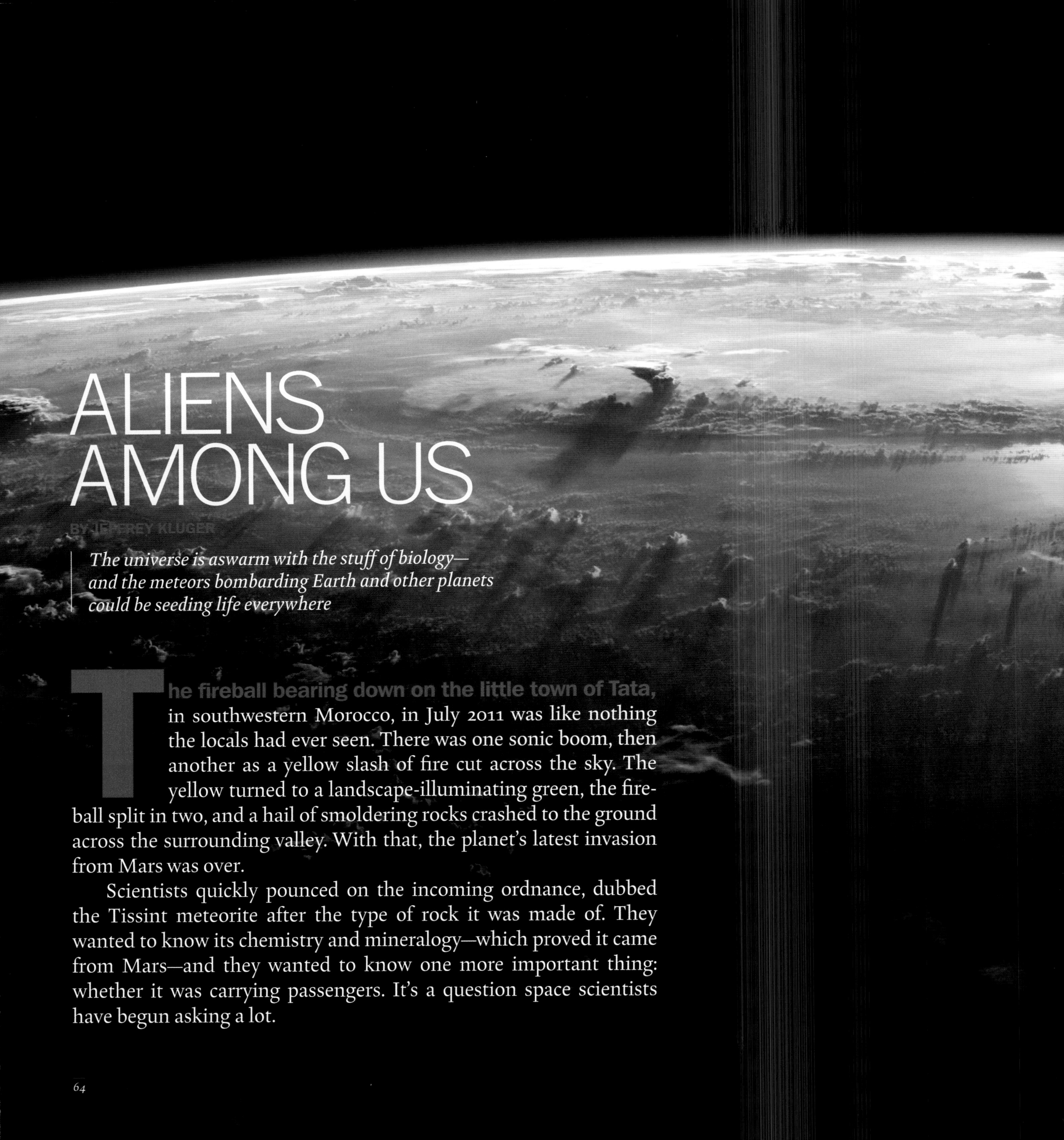

ALIENS AMONG US

BY JEFFREY KLUGER

The universe is aswarm with the stuff of biology—and the meteors bombarding Earth and other planets could be seeding life everywhere

The fireball bearing down on the little town of Tata, in southwestern Morocco, in July 2011 was like nothing the locals had ever seen. There was one sonic boom, then another as a yellow slash of fire cut across the sky. The yellow turned to a landscape-illuminating green, the fireball split in two, and a hail of smoldering rocks crashed to the ground across the surrounding valley. With that, the planet's latest invasion from Mars was over.

Scientists quickly pounced on the incoming ordnance, dubbed the Tissint meteorite after the type of rock it was made of. They wanted to know its chemistry and mineralogy—which proved it came from Mars—and they wanted to know one more important thing: whether it was carrying passengers. It's a question space scientists have begun asking a lot.

THROUGH THE LENS OF AN ASTRONAUT
This is a view of Earth's horizon as the sun sets over the Pacific Ocean. Anvil tops of thunderclouds are visible as well.

Precious Cargo

New computer simulations show how prebiotic material or microbial life could have originated in a distant solar system, then hitched a ride to Earth.

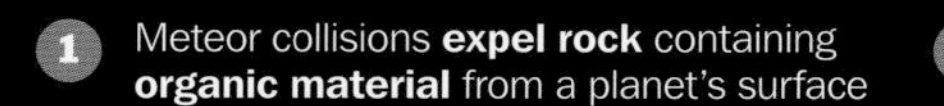

1. Meteor collisions **expel rock** containing **organic material** from a planet's surface
2. After **escaping** the planet's **gravitational pull**, the rock **drifts** through its solar system
3. Once at **the edge** of the solar system, the rock requires only **a flutter in its trajectory** to enter deep space

Life, as far as we can prove, exists only on Earth. There is our modest planet circling our modest star, and then there is the unimaginable hugeness beyond. Yet in that whole great cosmic sweep, we're the only little koi pond in which anything is stirring. That, at least, has been the limit of our science. But that limit is changing fast.

The cosmos, as scientists now know, is awash in the stuff of biology. Water molecules drift everywhere in interstellar space. Hydrogen, carbon, methane, amino acids—the entire organic-chemistry set—swirl through star systems and dust planets and moons. In 2009, NASA's Stardust mission found the amino acid glycine in the comet Wild 2. In 2003, radio telescopes spotted glycine in regions of star formation within the Milky Way. And meteors that landed on Earth have been found to contain amino acids, nucleobases—which help form DNA and RNA—and even sugars.

That raises a tantalizing question: If the building blocks of life can rain down anywhere, why not life itself—at least in the form of bacteria? Such an improbable idea—dubbed panspermia—has been chattered about by scientists since the 19th century. But back then, there wasn't much knowledge of what the cosmic ingredients of life would be or how to detect them even if they could be identified. That's all changed. A welter of new studies in the past few years have shed light on the panspermia idea—and in the process have changed our very sense of our place in the cosmos. Never mind the old image of life on Earth existing in a sort of terrestrial bell jar, sealed off from the rest of the universe. Our planet—indeed all planets—may be more like a great meadow, open to whatever spores or seedlings blow by.

"I think there's definitely a role meteorites have to play in at least getting prebiological materials to planets," says Chris Herd, a meteorite expert at the University of Alberta who has studied the Tissint rocks. "A lot has to go right for an actual microorganism to go from planet to planet. But in some cases, they just might survive the trip." If they made that trip to the ancient Earth, we may not merely have encountered aliens; we may be the aliens.

Martian Misfire

The search for life in rocks from space has not always been smooth. On Aug. 6, 1996, NASA stunned the world with a midday press conference announcing that a meteorite from Mars, prosaically known as ALH84001, contained evidence of what appeared to be fossilized bacteria.

Life on Mars, the headlines screamed—including one in TIME—and that was exactly the conclusion the researchers had tentatively reached. "It's an unbelievable day," said then NASA administrator Daniel Goldin. "It took my breath away." President Clinton, campaigning for re-election, took a break to weigh in too. "If this discovery is confirmed," he said in a White House statement, "it will surely be one of the most stunning insights into our universe that science has ever uncovered."

Stunning, yes, but that confirmation never came. Further study of 84001 failed to rule out inorganic processes for the seemingly biological clues it contained, and while the rock continues to spark debate, no one disputes that the evidence was not the slam dunk it seemed to be.

In the years since, the research has proceeded apace, even if the press releases have been more measured, and the case for panspermia is being convincingly rebuilt. Last year, Herd and his co-authors published a paper in the journal *Science* showing not just how biological material could get to Earth but also how it could survive a long trip in space.

The study focused on what's known as the Tagish meteorite, after the frozen lake in British Columbia on which it smashed itself to fragments on Jan. 18, 2000. Within days of the impact, scientists collected the debris—making no direct hand contact with it in order to prevent biological contamination—and put it in cold storage. When Herd and his colleagues got hold of four of the fragments and cracked them open, they found that the debris very much warranted such caretaking.

Distributed throughout the rock were not just the organics that had been seen before but also organics in different stages of sophistication, with simpler molecules giving way to complex ones and more complex ones still—a bit like finding caterpillars, cocoons and butterflies all in the same little nest. The rock, it seemed, had been acting as a sort of free-flying incubator, with traces of water trapped in its matrix combining with heat from radioactive elements to keep things warm and effectively pulsing.

"These asteroids form in space, you dump in organic molecules, a little water ice and a little heat, and then they just start to stew," says Herd. That slow cooking went on for millions of years until the heat and water eventually were exhausted, and the process shut down.

4 **Trace water** and **radioactive heat** within the rock **incubate its cargo** during the long journey

5 The **star cluster** in which our sun was born was once **tightly grouped**, reducing the rock's **transit time** as it is pulled into our solar system

6 The rock and its cargo, attracted by Earth's gravity, **plummet** through the atmosphere. If the organic material survives the plunge, it finds a very hospitable **new home**

This doesn't have to mean that similar rocks landing on Earth billions of years ago were the start of all terrestrial life—or even that they contributed to biological processes already under way. And yet the organics in the Tagish meteorite have a curiously familiar feature. Amino acids come in one of two varieties: left-handed and right-handed, defined by an asymmetrical structure that points either one way or the other. All earthly life uses the left-handed kind—a puzzle, since right-handed amino acids should work just as well—and the Tagish amino acids are left-handed too. Somehow, that southpaw bias got started on Earth. Herd's findings at least suggest that the influence could have come from beyond.

Cosmic Stowaways

It's easy enough to imagine how a meteor that accreted in space and spent its life flying could eventually find its way into the gravity field of a planet, if it came too close. Harder to figure is what it takes to get biologically contaminated material from the surface of one planet to another. Something, after all, has to launch the stuff in the first place. Typically that something is a meteor strike that hurls debris into space, where it slowly drifts from one world to the next. Earth and Mars have exchanged material this way for billions of years, though more in the early days of the solar system, when the cosmic bombardment was greater.

The kind of life that can get started on the warm, wet surface of a planet, contaminate its rocks and hitch a ride to the world next door is a lot more complex than the mere prebiology that can get cooked up in space. Most of those organisms—probably the single-celled kind like those the ALH84001 scientists thought they found—couldn't live through the shock of heating that occurs when debris is blasted into space, but the ones deep within the rock might. Surviving the hundreds of thousands or millions of years it would take to travel from world to world would not be impossible. Earthly bacteria that live in extreme environments may go dormant or even freeze-dry until conditions improve and they stir to life again.

In June 2012, investigators from the University of Colorado at Boulder studied bacteria found in the Atacama region of South America, where rain almost never falls and temperatures go from 13°F (–11°C) at night to 133°F (56°C) the next day. Microbes nonetheless thrive there, sucking energy from traces of carbon monoxide in the air and extracting moisture from exceedingly rare snowfalls. The rest of the time they hibernate. There's no reason an adaptation that nifty should be confined to earthly life.

Whatever biology is flitting about out there would not even have to be limited to traveling from planet to planet; it could also hop from solar system to solar system. This idea, known as lithopanspermia, was long considered impossible. Not only would the transit times between solar systems be prohibitively long for even the hardiest bacteria—on the order of 1.5 billion years—but the speed a space rock needs to travel to escape the gravity of its home solar system is too great for it to be captured by another. In September 2012, however, a team of researchers from Princeton University, the University of Arizona and the Centro de Astrobiología in Spain figured out a neat solution that sidesteps these problems.

Most lithopanspermia models assumed that the only way a rock could escape a solar system was if it passed too close to a large body like Jupiter and was gravitationally ejected at a speed of about 18,000 mph (29,000 km/h). But the investigators in the recent study used a computer to model a slow-boat escape known as weak transfer, in which a rock gradually drifts out through a solar system until it's so far from its parent sun that the slightest flutter in its trajectory could tip it into interstellar space.

"At this point," says Princeton astrophysicist Edward Belbruno, one of the co-authors, "mere randomness determines whether it gets out or not." And never mind the extreme distances to the nearest solar systems. About 4.5 billion years ago, the infant sun was part of a tight grouping of nascent stars known as the local cluster. The herd dispersed after less than 300 million years, but a weak-transfer rock that escaped within that window could have reached the next solar system in about a million years. "Trillions of rocks could escape a solar system," says Belbruno. "Over the course of 300 million years, about 3 billion might have struck Earth."

It's impossible to know if even one of those 3 billion would have harbored biological material, especially so early in the history of the local stars. But if the new studies say anything, it's that it's equally impossible to continue to see the Earth and its organisms as somehow separate from the rest of the cosmos. The building blocks of biology are everywhere; life, it seems increasingly likely, could be too.

THE SEARCH FOR NASA'S MISSION

BY DANIEL CRA[illegible]

Despite its successful Mars rovers, the once high-flying agency has been brought down to Earth by budget cuts and uncertainty about its next big act

There isn't much wind in the Gale crater on Mars. But as word came to NASA that one of its wildest interplanetary endeavors ever—the Jet Propulsion Laboratory's hovering "sky crane"—had successfully deposited the SUV-size Curiosity rover onto the crater's rugged surface, you could almost feel a breeze hoisting the space agency's beleaguered sails. From Pathfinder's tiny Sojourner to the heart-tugging twins Spirit and Opportunity, the rover fleet exploring Mars has captured the public's imagination better than anything in NASA's past two decades

NASA'S DREAM CHASER, *a rendering of the shuttle under development by Sierra Nevada Corp., to ferry cargo and crew between Earth and the International Space Station*

THE ORION MULTI-PURPOSE CREW *vehicle, set to launch in 2017, which will carry a crew of four beyond low Earth orbit—to the moon, an asteroid and perhaps to Mars. Here, NASA technicians test its acoustics.*

of space endeavors. Now brash Curiosity takes a turn, with ambitious plans to identify the chemical components forming the crater's dusty soil. The rover has already found simple organics, meaning there's a chance planetary scientists may at last learn whether Mars harbors the organic compounds that serve as the molecular building blocks for life on Earth.

The successful touchdown comes at something of a desperate hour, since NASA's unmanned interplanetary programs are under duress, facing proposed budget cuts of about $300 million primarily due to cost overruns on the James Webb Space Telescope. More than $200 million of the cuts have gouged the $587 million Mars exploration program, home of the beloved rovers. Already the budget shortage has placed a sample-and-return mission in jeopardy and left NASA in the egg-faced position of backing out of a prior commitment to conduct two robotic Mars missions with the European Space Agency.

NASA ASTRONAUT *Rex Walheim participates in an evaluation of the Advanced Crew Escape Suit (ACES) in the Active Response Gravity Offload System (ARGOS) at Johnson Space Center in Houston. The ARGOS system is intended to simulate in one system the reduced gravity effects of the moon, Mars, and space.*

Manned spaceflight is facing even bigger hurdles. There is little bipartisan support for increasing spending to the levels it would require. Constellation, the dual-rocket program intended to succeed the space shuttles and carry astronauts to the moon, was canceled in favor of a single-rocket Space Launch System that is cheaper but still criticized as too expensive for its more modest capabilities. With NASA on a seemingly aimless path, international interest in pooling resources for manned spaceflight efforts beyond the International Space Station has eroded.

At home, NASA's supporters have fractured into two partisan groups: one that endorses the space agency's post-shuttle effort to privatize space launches and another that feels U.S. spaceflight capability should not be left to a handful of unproven, under-regulated upstarts. NASA's flat budget isn't helping either cause. In the agency's Cold War heyday, federal spending for the space program amounted to more than 4% of the federal budget; today's $17.8 billion allotment checks in at just .48%, its lowest point since 1960, and sequestration cuts another $900 million from its budget.

Bad as it seems, it's not NASA's first trip through a financial asteroid belt. The agency has survived lean years in the past, and it spent most of the 1970s similarly grounded. Public and political concern about expenses isn't exactly new either; during the headstrong Apollo years, polls indicated a majority of Americans considered the space program too costly. Still, today's hard times are raising questions about where NASA is headed and what will be left of it once it finally gets there. "Back when they were starting in on the shuttle program, we had just returned from the moon, and these guys were heroes," says Roger Launius, senior curator for space history at the National Air and Space Museum. "That's not the case anymore."

Lowered cachet equals less political clout. President Obama snuffed hopes for a new moon mission and arguably wasted three years of Constellation development after a blue-ribbon panel issued a Yoda-like recommendation: Either do or do not, there is no try. Translation: If the U.S. couldn't commit the necessary billions, there was no point spending for half a program. With the economy a mess, Obama scrubbed the mission.

In so doing, he ushered forth a long-discussed plan to turn over all near-Earth-orbit flights to private industry. However defensible the decision in view of the economy, it led critics to accuse him of stripping NASA of its space access. NASA administrator Charles Bolden, however, insists that's revisionist history. "This started with the past administration, in 2004, when a Congressional mandate said to phase out the shuttle over the next 10 years and bring about a commercial capability, but provided no funds. The present President took a risk and put funds toward it."

That risk has produced an early payoff with SpaceX. The Hawthorne, Calif., company's promising success with its Falcon 9 rocket and companion Dragon capsule has quickly become what John Holdren, director of the White House Office of Science and Technology Policy, describes as "a cornerstone of the President's plan for maintaining America's leadership in space." Both he and Bolden say private-sector involvement frees up NASA's resources to tackle more demanding challenges. SpaceX has

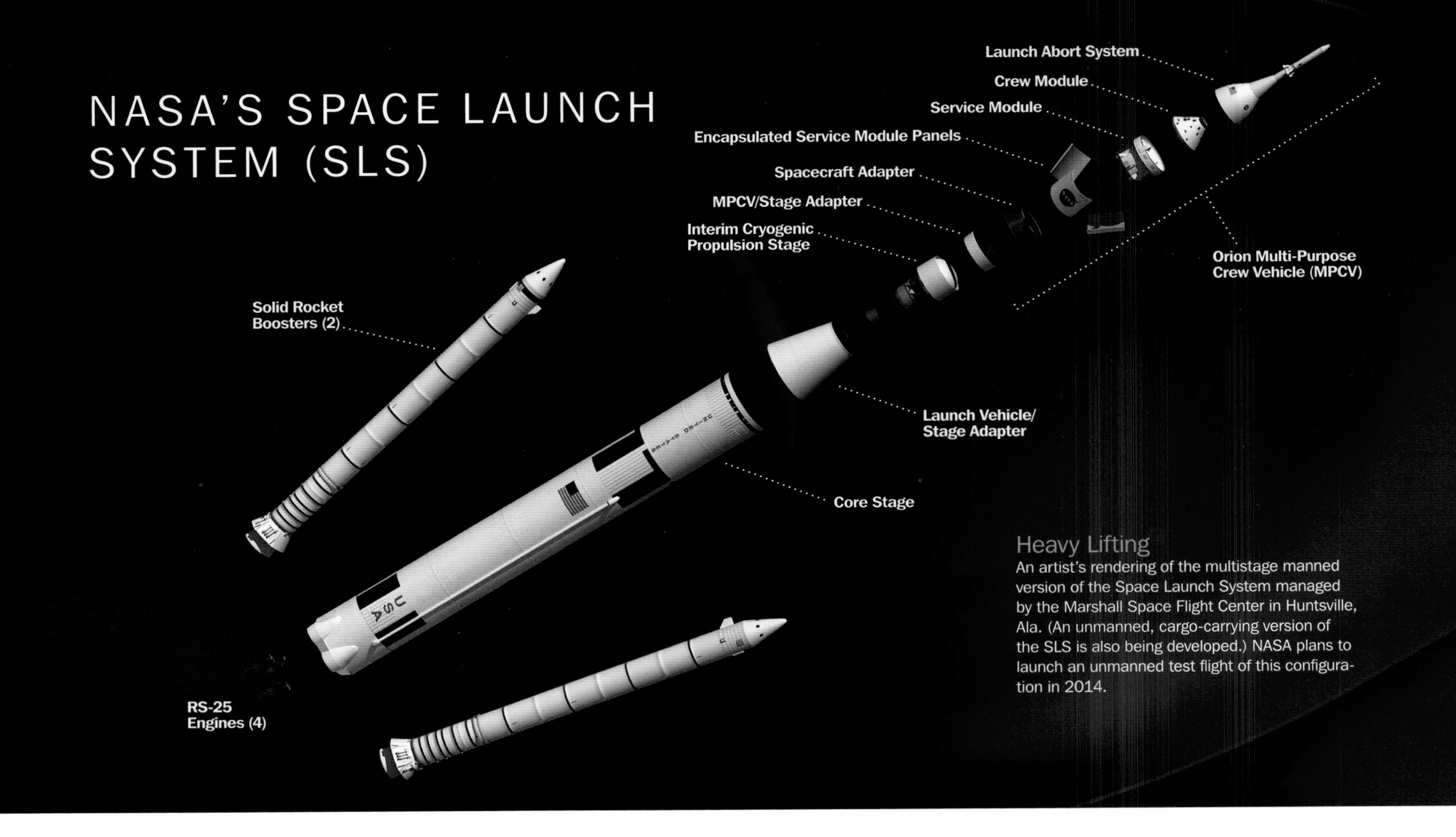

Heavy Lifting
An artist's rendering of the multistage manned version of the Space Launch System managed by the Marshall Space Flight Center in Huntsville, Ala. (An unmanned, cargo-carrying version of the SLS is also being developed.) NASA plans to launch an unmanned test flight of this configuration in 2014.

"flung the door wide open," Bolden says, "allowing us to think about Mars and lunar exploration and asteroids."

The strategic shift masks a surprising truth: Unless you had a role in the shuttle's launch or worked at the Kennedy Space Center, you'd find little that has changed in NASA's current day-to-day operation. Roughly 85 cents of every NASA dollar has gone to outside industry for decades, so relying on private enterprise to ferry cargo and eventually astronauts to the space station merely evolves a longstanding practice. "Apollo was contracted, shuttle was contracted, the mission teams were contracted," says Launius. "When SpaceX lifted off, it was virtually the same as every other NASA launch."

There's a big asterisk, however. Historically, the ability to develop and acquire its own hardware is what separated NASA from other government agencies. Now the rockets and capsules belong to private enterprises. Unless SpaceX emblazons NASA's logo on Dragon's exterior like a NASCAR entry, the agency's participation won't be obvious. That, not the outsourcing itself, is the most significant change to hit the space agency. And from an image standpoint, it's no small difference. As SpaceX, Orbital Sciences, and several other private launch companies arise, their success seems poised to feed the public perception of NASA as little more than a glorified grant-application center.

Bolden rejects that notion, insisting that with the private sector handling low Earth orbit, the space agency's own hardware and astronauts will be flying the bigger off-planet missions to . . . wherever. "We don't have to worry about getting people into low Earth orbit anymore, so we can take time to start developing vehicles that are going to get us beyond," he says. NASA's 2013 budget includes $1.9 billion (a $326 million cut) for the Orion Multi-Purpose Crew vehicle and the Space Launch System. Assuming Congress continues budgeting the program, the first SLS test flight is slated for 2017.

So where does all this leave NASA? Aside from the manned-spaceflight program, NASA's efforts should survive the usual budgetary ebb and flow, especially since the Webb telescope overruns amount to a one-time expense. Most analysts feel the unmanned robotics programs, though hampered in 2013, will eventually return at full strength, since planetary scientists have clear priorities with bipartisan Congressional and public support.

Only the manned-spaceflight program is under genuine long-term threat, and real solutions likely won't arrive until NASA finds a destination with bipartisan appeal. As any armchair NASA administrator will tell you, kick-starting the program isn't, well, rocket science—as long as you have an obvious destination. NASA hasn't had that luxury since 1972. "What is lacking with NASA today is a clear sense that it has a credible mission in human space exploration that would support the development of new, post-shuttle capabilities," says Scott Pace, director of Washington University's Space Policy Institute. His reasoning: NASA is a mission-

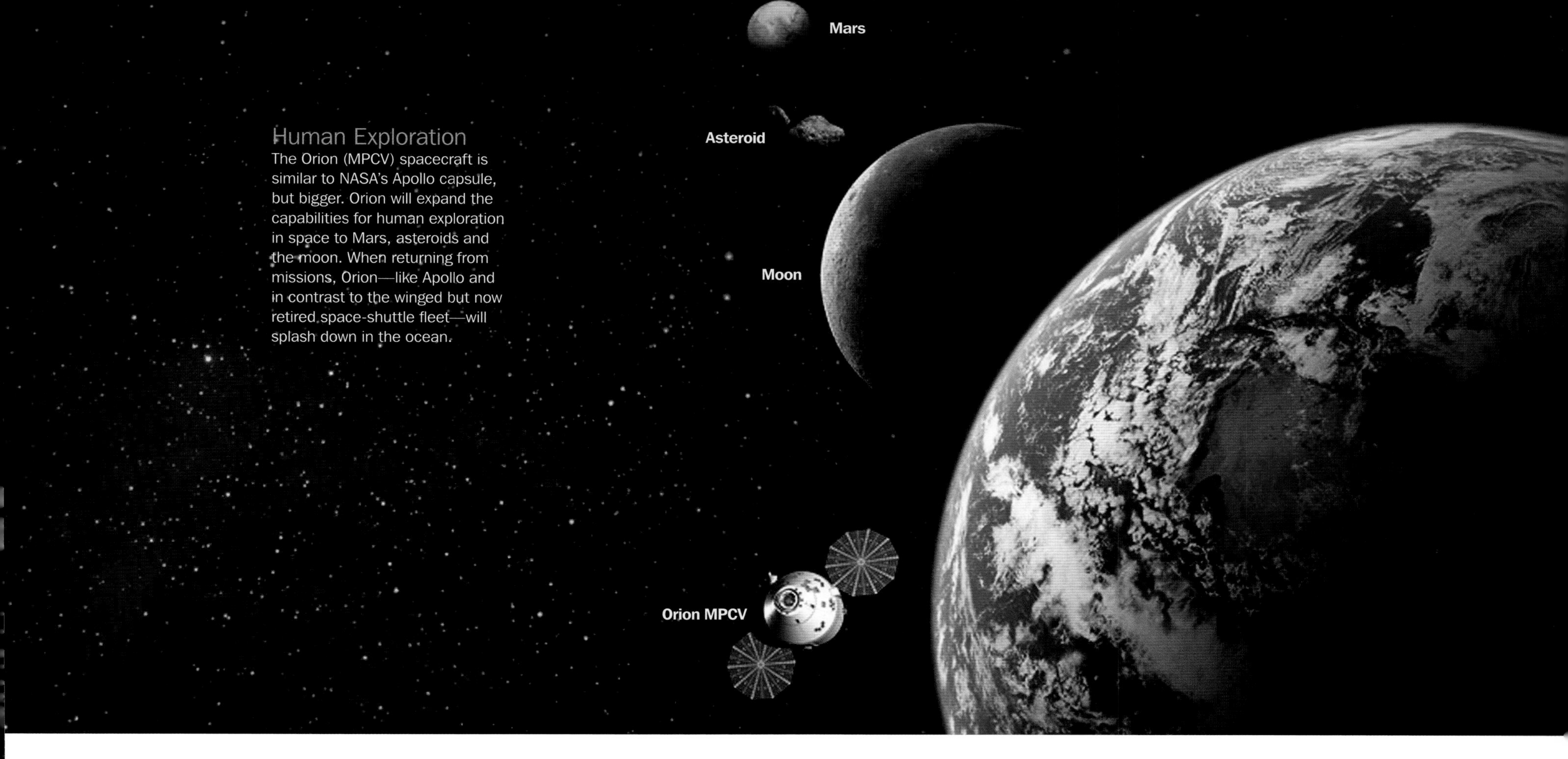

Human Exploration

The Orion (MPCV) spacecraft is similar to NASA's Apollo capsule, but bigger. Orion will expand the capabilities for human exploration in space to Mars, asteroids and the moon. When returning from missions, Orion—like Apollo and in contrast to the winged but now retired space-shuttle fleet—will splash down in the ocean.

driven agency that needs clear goals, strong leadership, and bipartisan support to be successful. The International Space Station, though an immense technological achievement, is not a solution. As Launius puts it, "Exploration means actually going somewhere." Between the shuttle and the space station, NASA has traveled in circles for more than 30 years. Though the missions succeeded in making near-Earth orbit a routine destination, they added up to an Apollo sequel that was good but not great.

Mars doesn't offer an answer either, at least not for a while. The journey itself would take about seven months, followed by an 18-month wait for Earth and Mars to reach the orbit locations necessary for the return trip. That means three years away from Earth in cramped conditions that would make the cluttered space station feel like a luxury resort. Psychological hurdles aside, the astronauts face prolonged exposure to deadly cosmic rays and a low-gravity environment—and brace yourself for the cost: In 1989 NASA estimated that sending humans to Mars would require $400 billion, which inflates to $750 billion today.

Visiting an asteroid might offer an intermediate target, but detractors point to the added challenges of visiting a zero-gravity, zero-atmosphere environment and to its limited scientific benefit. Perhaps a more viable plan is a decidedly unsexy trip to the Lagrange points, positions in space where Earth's pull is fully balanced by gravity from either the sun or the moon. A spacecraft could thus remain for long periods without consuming much energy, an opportunity to study the effects of sending astronauts away from Earth's magnetosphere. Then again, "it's hard to get people behind something that isn't there," Launius says. "People want to see astronauts actually step onto something, not just float at some point in space."

Many feel the solution had already been proposed and was under way until Obama scuttled it: a new moon mission. Pace (who openly supported Obama's opponent in the 2012 election) suggests the organizations and skills that enabled us to operate on the lunar surface need to be rebuilt before NASA can give serious consideration to sending humans to Mars. "It's a matter of rebuilding intellectual capabilities," Pace says, "and training a new generation of engineers, scientists, and technicians."

It is also, he suggests, a reflection of what society values: "We reached the moon, built and operated a shuttle fleet for 30 years, explored the solar system, and built an International Space Station for less than the cost of the 2009 Recovery Act." In other words, people find money for causes they truly support. Astrophysicist Neil deGrasse Tyson, stumping to lift NASA's budget, pointedly informed Congress that the only reason NASA reached the moon was because the U.S. was "at war" with the Soviet Union. "When you perceive your security to be at risk," Tyson said, "money flows like rivers." But the Cold War is long over, and NASA's chances of launching its own manned rocket likely depend on whether the nation can agree on a destination. If so, NASA and its long-term off-planet aspirations stand a reasonable chance. Otherwise, U.S. spaceflight may remain an uneasy public-private alliance, headed up but never away.

差一点点成功

THEY'RE LAUNCHING ALL OVER THE WORLD

BY DANIEL CRAY

More than a dozen space programs in China, India and other countries have ambitious plans, but navigating the global economy may be as daunting as escaping Earth's orbit

THE LAST TIME AMERICA'S SPACEFLIGHT PROGRAM LAGGED BEHIND ANOTHER NATION'S WAS in 1961, and Yuri Gagarin was peering down at us as he circled Earth in the cramped, 6-ft.-diameter confines of spherical Vostok 1. Eight years later, human spaceflight had became a world-leading skill set that, while not unique to the U.S., seemed as entrenched in our national identity as Neil Armstrong's lunar bootprint.

Even now, with the U.S. hitching rides from the Russians and slashing interplanetary science budgets, there is no indication that other nations are eying the situation as an opportunity to leapfrog over America's accomplishments with, say, a manned trip to Mars. "No one's itching to race," says Dwayne Day, a space-policy analyst at the National Research Council. "The United States still outspends everybody else and is still involved in the broadest range of spaceflight activities."

But other nations, most notably China, have ambitious programs of their own. By moving steadily forward with proven technologies, the Chinese have quietly assembled a program many analysts equate to NASA's Gemini years. In June 2012, China became the third nation to successfully dock a spacecraft in orbit when Shenzhou 9 carried three "taikonauts" to the country's Tiangong-1 space lab. "The American public isn't patient; but for the Chinese, with a 5,000-year history, it's real easy for them to take 30 years," says Joan Johnson-Freese, a professor of national security affairs at the U.S. Naval War College in Newport, R.I. Whereas Apollo launched often with small advances, she says, "the Chinese model is, launch every two years and take a big step."

In December 2011 the Chinese government unveiled a five-year plan for a new orbiting lab, a

CHINA'S LONG MARCH 2F *carrier rocket, one of the country's most successful, is assembled at the Jiuquan Satellite Launch Center in northwest China.*

more powerful manned spacecraft, and an unmanned sample-and-return mission to the moon. They also hope to improve their global navigation satellite system and, longer term, to assemble a 60-ton Skylab-size space station. Their Long March 5 rocket, already in development, is designed to lift about 25 tons to low Earth orbit—not exactly a match for the Apollo-era Saturn V but comparable to the current Delta IV rocket used by the U.S.

While increasing gaps between China's launches may hint at economic roadblocks, analyst Day believes the Chinese have established "a pace that suits their needs." Still, the Long March 5 is delayed until 2015, and expenses are reportedly mounting. Johnson-Freese wonders whether sporadic reports that China hopes to mine the moon for rare helium-3,

THE SHENZHOU-9 *successfully docked at the Tiangong-1 space lab in June 2012, a first step in China's plan to build a space station by 2020. The crew (from left): Liu Yang, China's first female taikonaut; Jing Haipeng, mission commander; and Liu Wang*

sought for nuclear-fusion research, suggest a political demand for justification of the costs of the space program. "That's been used by every country in the world," she says. "Whenever politicians ask why we should go to the moon, you pull out the helium-3." Though the Chinese have been surprisingly open about discussing their spaceflight plans at conferences and on the Internet, they remain weighed down by authoritarian bureaucracy and secrecy requirements.

But China's growing list of space achievements appears to carry political weight with at least one of its neighbors. India, which has long focused on satellite systems for communications and weather forecasting, has recently disclosed plans for human spaceflight and an ambitious moon-mapping mission, with target dates that would potentially beat the Chinese. "All of a sudden India is feeling the pressure of prestige," Johnson-Freese says. "China is the regional space leader, and India doesn't like where that leaves them."

While China and perhaps India are marching steadily forward, some analysts say Russia is standing in place. The country with a proud space history now seems focused on an aging breadwinner—the single-use Soyuz capsule—at the expense of pioneering anything new. "They issue announcements that they're developing new spacecraft, but you just cannot believe them," Day says. "The money that has started to flow back into their space program is clearly replenishing their military capabilities first."

Despite the fact that money for space exploration is tight worldwide, space economics are currently good for Russia, which is raking in billions by selling no-frills rides on their 1970s-era capsule. They receive $63 million for every U.S. trip to the International Space Station, a price the U.S. can't haggle over because there is no reliable alternative. Overall, NASA says Russia's space agency will receive $753 million to ferry 12 U.S. astronauts to the ISS in 2014 to 2016. Since 2007, the Russians have reportedly made about $2.5 billion from NASA and its partner agencies.

Alas, that newfound cash pile has coincided with a string of troubles, including the 2011 crash of a Soyuz rocket and the loss of a $170 million unmanned Martian space probe that wound up stranded in Earth orbit. Analysts feel the Russian space program is having trouble transitioning to a younger generation of engineers, scientists and technicians. "They've had a tremendous brain drain," says Day. "Russian space scientists left in droves in the 1990s."

Many departed for the European Space Agency and its extensive but budget-challenged interplanetary robotics program. Japan, by contrast, embraced a small-budget agenda. The Japanese Kaguya spacecraft mapped the moon, and the Hayabusa asteroid probe successfully returned soil samples to Earth, both at relatively low cost. "Financially speaking, Japan has gotten to space on a shoestring, less than $4 billion per year," says Saadia Pekkanen, a political scientist specializing in Asian space policy at the University of Washington. "Their real issue is human capital, the biggest limiting factor for Japan as it attempts to move forward." Japan has a relatively modest space sector of only 8,400 people. China, by comparison, employs an estimated 48,000. Many analysts feel Japan won't be a major player in space until its government boosts its commitment of human and financial resources.

That's the thread linking all government space programs: the challenge of navigating a difficult global economy. Many of the world's 14 major space agencies have launch vehicles and interplanetary robotics, but all of them expect the U.S. to bring its human-spaceflight expertise to the table. "Until we sort things out," says Scott Pace, director of George Washington University's Space Policy Institute, "we will be less influential in the international space community." How quickly that happens may be the only space race that matters.

RUSSIAN SPACE AGENCY PERSONNEL *inspect the Soyuz TMA-22 capsule that landed in Kazhakhstan in April 2012 after ferrying home its joint U.S.-Soviet crew from a five-month mission on the International Space Station.*

CREW MEMBERS OF SOYUZ TMA-04M, *the joint U.S.-Soviet science team, gather on camera after docking at the International Space Station. The mission launched from the Baikonur Cosmodrome in Kazakhstan.*

A HALF-CENTURY IN SPACE

MANNED MILESTONES

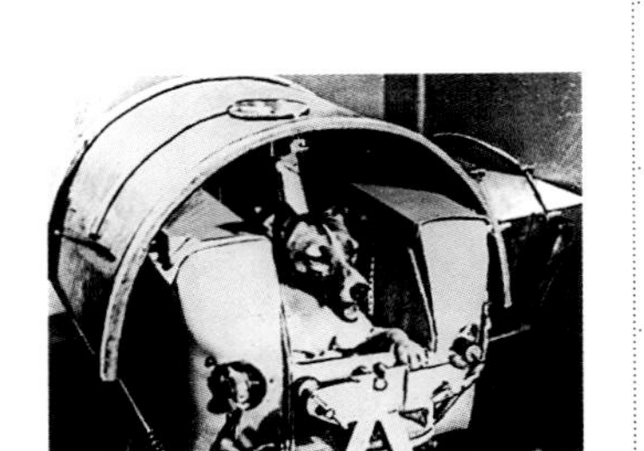

NOVEMBER 3, 1957
Sputnik 2 carries the dog Laika into space as test animal

APRIL 9, 1958
NASA names seven pilots to the original U.S. astronaut corps

JANUARY 31, 1961
U.S. sends chimpanzee Ham into space to pave way for humans

APRIL 12, 1961
First man in space: Soviet cosmonaut Yuri Gagarin

MAY 5, 1961
U.S. launches its first astronaut, Alan Shepard

MAY 25, 1961
President John F. Kennedy declares goal of landing a man on the moon

FEBRUARY 20, 1962
John Glenn orbits Earth

JUNE 16, 1963
First woman in space: Soviet cosmonaut Valentina Tereshkova

MARCH 18, 1965
Cosmonaut Aleksei Leonov walks in space

JUNE 3, 1965
Ed White performs first U.S. space walk

JANUARY 27, 1967
Ed White, Gus Grissom and Roger Chafee die in Apollo 1 preflight test

DECEMBER 24, 1968
Apollo 8 makes first manned orbit of the moon

JULY 20, 1969
Apollo 11 moon landing. Historic footsteps of Neil Armstrong, Buzz Aldrin, carried on live TV

APRIL 13, 1970
Explosion forces Apollo 13 to abort mission and make emergency return to Earth

1957 · 1960 · 1963 · 1966 · 1969

UNMANNED MILESTONES

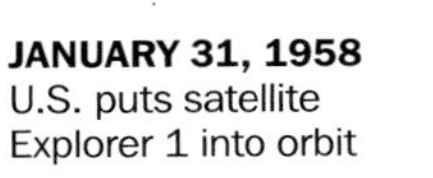

JANUARY 31, 1958
U.S. puts satellite Explorer 1 into orbit

OCTOBER 4, 1957
Soviets launch Sputnik 1, first man-made object to orbit Earth

APRIL 1, 1960
U.S. launches TIROS-1, the debut of an ongoing series of weather satellites

JULY 12, 1962
Telstar 1 satellite relays transatlantic TV signal from U.S. to France

JULY 22, 1962
First of 10 Mariner missions (three fail) to Mars, Venus and Mercury

JULY 14, 1965
Mariner 4 takes first photos of another planet, Mars

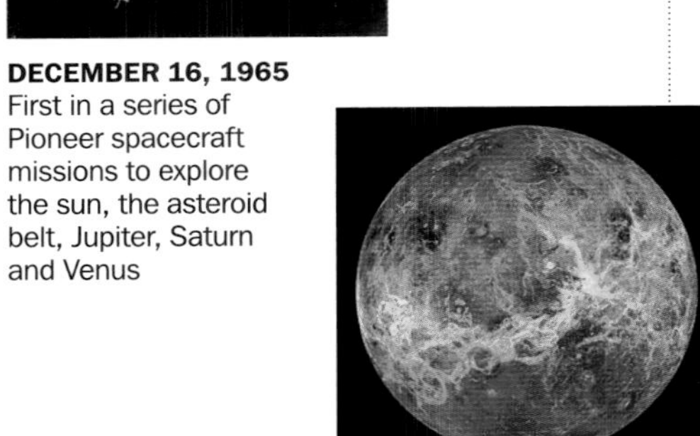

DECEMBER 16, 1965
First in a series of Pioneer spacecraft missions to explore the sun, the asteroid belt, Jupiter, Saturn and Venus

DECEMBER 15, 1970
The Venera 7 (one of many Soviet missions to Venus) makes first soft landing on another planet

APRIL 19, 1971
Soviets launch first space station, Salyut 1

JULY 31, 1971
Debut of lunar rover vehicle dubbed the "moon buggy"

MAY 14, 1973
U.S. launches Skylab; six years later, its fall to Earth is NASA embarrassment

JULY 17, 1975
U.S. and U.S.S.R. spacecraft dock; crews work together for two days

APRIL 12, 1981
Columbia is launched on U.S. space shuttle's first mission

JUNE 18, 1983
Sally Ride is first U.S. woman in space. (Ride died of cancer in July 2012)

NOVEMBER 28, 1983
First mission of orbiting Spacelab

1972 1975 1978 1981 1984

CONTINUED

APRIL 5, 1973
Pioneer 11 heads to Saturn to take first close-up images of planet and its rings

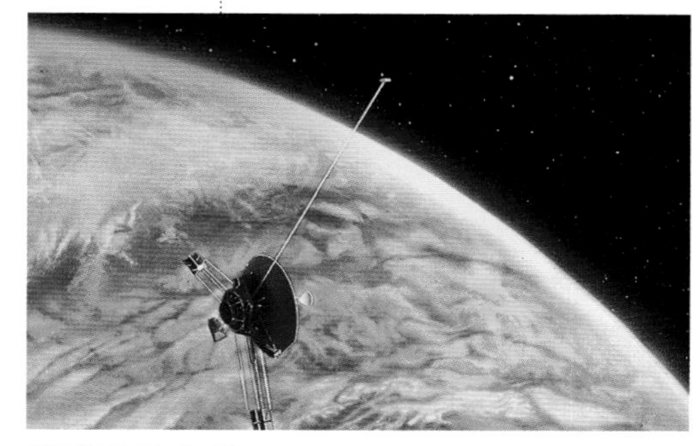

DECEMBER 3, 1973
Pioneer 10 flies over Jupiter's cloud tops

JULY 20, 1976
Viking 1 lands on Mars; Viking 2 arrives 45 days later

AUGUST–SEPTEMBER 1977
Voyagers 1 and 2 leave for Saturn and Jupiter and their moons; Voyager 2 goes on to Uranus and Neptune

MAY 20, 1978
Pioneer Venus orbiter launched to study planet's atmosphere and map its surface; Pioneer Venus Multiprobe follows on Aug. 8

DECEMBER 15 & 21, 1984
Soviet Vega missions set out for Venus; redirected to fly by Halley's comet in March 1986

JANUARY 28, 1986
Challenger explodes, killing seven, including Christa McAuliffe, first teacher in space

FEBRUARY 20,1986
Soviet space station Mir launched; crews occupy it for nearly 13 years

SEPTEMBER 29, 1988
Discovery, first shuttle launch since 1986 *Challenger* tragedy

DECEMBER 2, 1993
First of five shuttle missions (and 23 space walks) to repair Hubble telescope

JANUARY 8, 1994
Russian Valeri Polyakov begins record 438 days in space

JUNE 29, 1995
Atlantis docks with Russian space station Mir

DECEMBER 4, 1998
U.S. and Russia begin joint construction of International Space Station

1985 · · 1988 · · 1991 · · 1994 · · 1997 ·

CONTINUED FROM PREVIOUS PAGE

UNMANNED MILESTONES

MAY 4, 1989
Magellan spacecraft launched on mission to map Venus

OCTOBER 18, 1989
Galileo probe to Jupiter and its moons launches from space shuttle *Atlantis*

APRIL 24, 1990
Hubble Space Telescope deploys on mission to reveal the cosmos

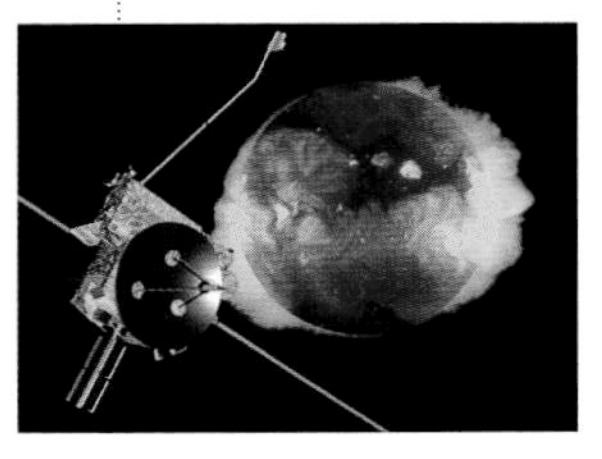

OCTOBER 6, 1990
Discovery launches solar probe Ulysses

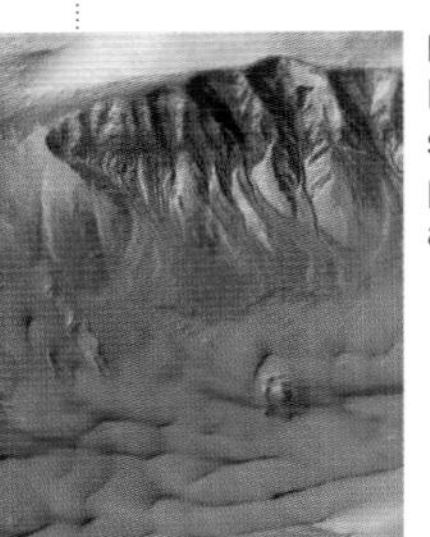

NOVEMBER 7, 1996
Mars Global Surveyor sets out; orbits the planet for nine years and finds signs of water

JULY 4, 1997
Pathfinder lands on Mars; rover Sojourner explores planet for three months

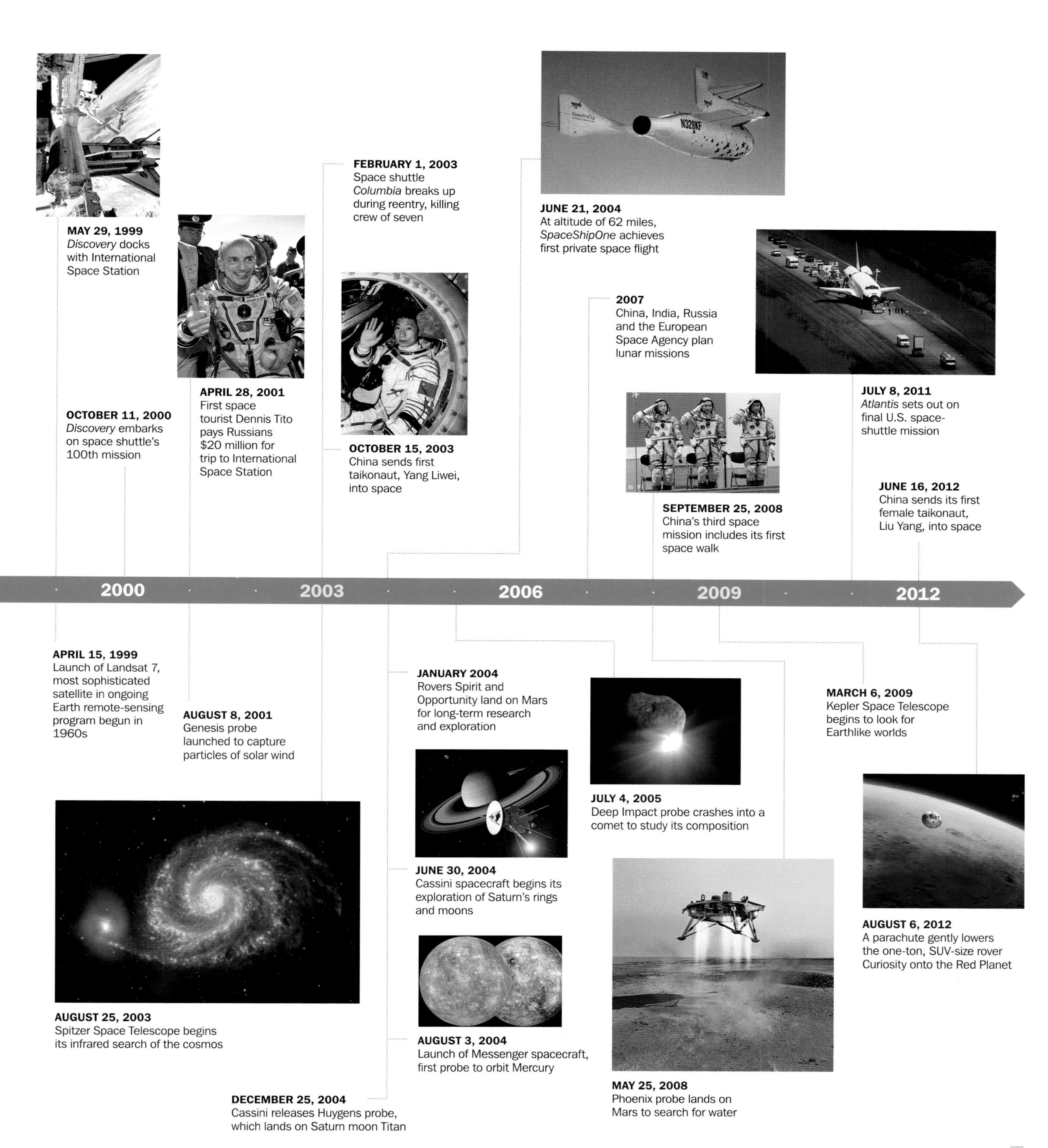

APRIL 15, 1999
Launch of Landsat 7, most sophisticated satellite in ongoing Earth remote-sensing program begun in 1960s
MAY 29, 1999
Discovery docks with International Space Station
OCTOBER 11, 2000
Discovery embarks on space shuttle's 100th mission
APRIL 28, 2001
First space tourist Dennis Tito pays Russians $20 million for trip to International Space Station
AUGUST 8, 2001
Genesis probe launched to capture particles of solar wind
FEBRUARY 1, 2003
Space shuttle Columbia breaks up during reentry, killing crew of seven
AUGUST 25, 2003
Spitzer Space Telescope begins its infrared search of the cosmos
OCTOBER 15, 2003
China sends first taikonaut, Yang Liwei, into space
JANUARY 2004
Rovers Spirit and Opportunity land on Mars for long-term research and exploration
JUNE 21, 2004
At altitude of 62 miles, SpaceShipOne achieves first private space flight
JUNE 30, 2004
Cassini spacecraft begins its exploration of Saturn's rings and moons
AUGUST 3, 2004
Launch of Messenger spacecraft, first probe to orbit Mercury
DECEMBER 25, 2004
Cassini releases Huygens probe, which lands on Saturn moon Titan
JULY 4, 2005
Deep Impact probe crashes into a comet to study its composition
2007
China, India, Russia and the European Space Agency plan lunar missions
MAY 25, 2008
Phoenix probe lands on Mars to search for water
SEPTEMBER 25, 2008
China's third space mission includes its first space walk
MARCH 6, 2009
Kepler Space Telescope begins to look for Earthlike worlds
JULY 8, 2011
Atlantis sets out on final U.S. space-shuttle mission
JUNE 16, 2012
China sends its first female taikonaut, Liu Yang, into space
AUGUST 6, 2012
A parachute gently lowers the one-ton, SUV-size rover Curiosity onto the Red Planet
2000
2003
2006
2009
2012
N328KF

NASA SCIENTISTS *field test the Lunar Electric Rover near Flagstaff, Ariz. The desert's dusty, rocky terrain and extreme temperature swings are the closest scientists can come to conditions on the moon and other potential planetary destinations.*

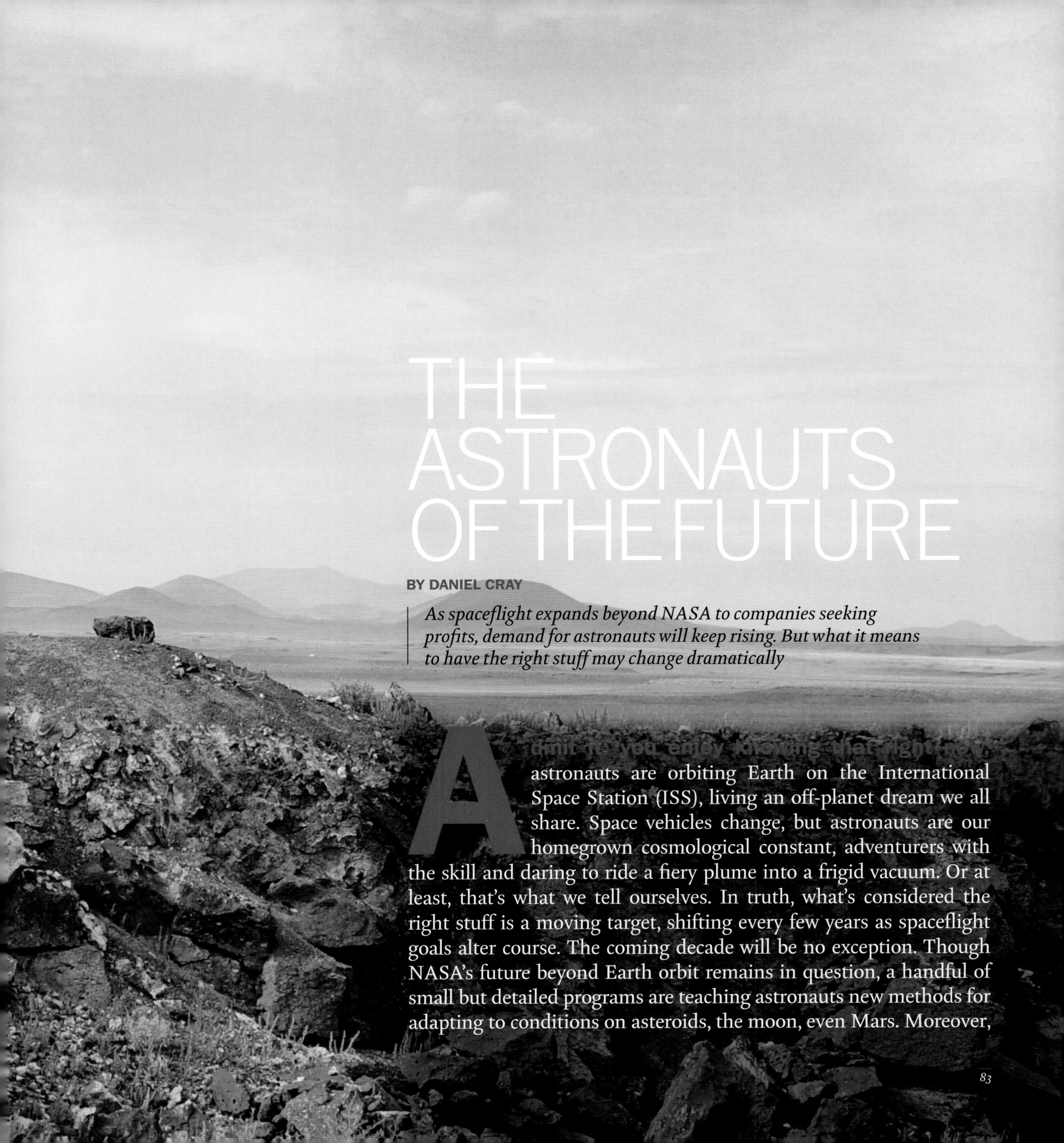

THE ASTRONAUTS OF THE FUTURE

BY DANIEL CRAY

As spaceflight expands beyond NASA to companies seeking profits, demand for astronauts will keep rising. But what it means to have the right stuff may change dramatically

Admit it: you enjoy knowing that right now, astronauts are orbiting Earth on the International Space Station (ISS), living an off-planet dream we all share. Space vehicles change, but astronauts are our homegrown cosmological constant, adventurers with the skill and daring to ride a fiery plume into a frigid vacuum. Or at least, that's what we tell ourselves. In truth, what's considered the right stuff is a moving target, shifting every few years as spaceflight goals alter course. The coming decade will be no exception. Though NASA's future beyond Earth orbit remains in question, a handful of small but detailed programs are teaching astronauts new methods for adapting to conditions on asteroids, the moon, even Mars. Moreover, with U.S. orbital spaceflight duties making a transition to commercial enterprise, many of the astronauts are shifting as well, learning new skills for a variety of different spacecraft. "Astronauts can't afford to be specialized anymore," says Michael Lopez-Alegria, who flew three shuttle missions and commanded an ISS crew. "You really have to do it all."

Astronaut selection and training remains NASA's exclusive domain, but that too may change. Already, a growing cadre of retired astronauts is being courted by commercial spaceflight companies. "I did not expect the mass exodus that we're seeing today," says Chris Ferguson, who piloted the final space shuttle mission and is now at Boeing, assisting in the development of the company's CST-100 space capsule. Only 54 astronauts remain on NASA's active-duty roster, down from a peak of about 150 in 2000. One reason: Veteran astronauts who joined NASA to fly the shuttle are looking elsewhere for opportunities. "The private sector's done a very good job of hiring former astronauts with the intent that they will fly," says Peggy Whitson, former chief of the Astronaut Office at Johnson Space Center. "Our challenge is keeping some of that experience base here within NASA."

The first commercial manned spaceflights may be staffed by private crews, in part because NASA feels it can't legally certify the safety of commercial vehicles until a late stage of development and therefore can't put NASA astronauts inside. As a result, "it's not yet clear whether company pilots will fly these commercial-crew vehicles or NASA pilots or some combination," says Lopez-Alegria, who is now president of the Commercial Spaceflight Federation, a private-spaceflight industry association founded in 2005. "But the test program for these things will probably be done by company pilots." Count Boeing pilots among them. "We'll do a shakedown flight, then the plan is to basically lease the vehicle to NASA," Ferguson says. "Once it's fit for operational flight, we'll turn it over to the NASA crews to operate."

There's also a chance NASA could join forces with the commercial effort even if companies decide to recruit their own astronaut teams. "I don't think we'd be opposed to training," Whitson says. "It's just a question of when would that happen and how." And if. Private companies may not be willing to spend the kind of money NASA has traditionally lavished on astronaut training. "Maintaining flying proficiency does not come cheaply," says Ferguson, who suggests the commercial airline industry, in which companies hire already-trained pilots and give them additional training for new aircraft and technology, is a better model.

There will be no shortage of candidates. The exodus of astronauts from NASA to commercial spaceflight companies apparently didn't concern the near-record 6,300 people who applied to join NASA's 2012 class of no more than 15 astronauts. That class will become only the second group training exclusively for ISS missions, a huge paradigm shift from the shuttle's tactical pilots and mission specialists. Each ISS astronaut, by contrast, must handle robotics, scientific payloads, station repairs and space walks—not to mention normal ISS operations. "The crew mem-

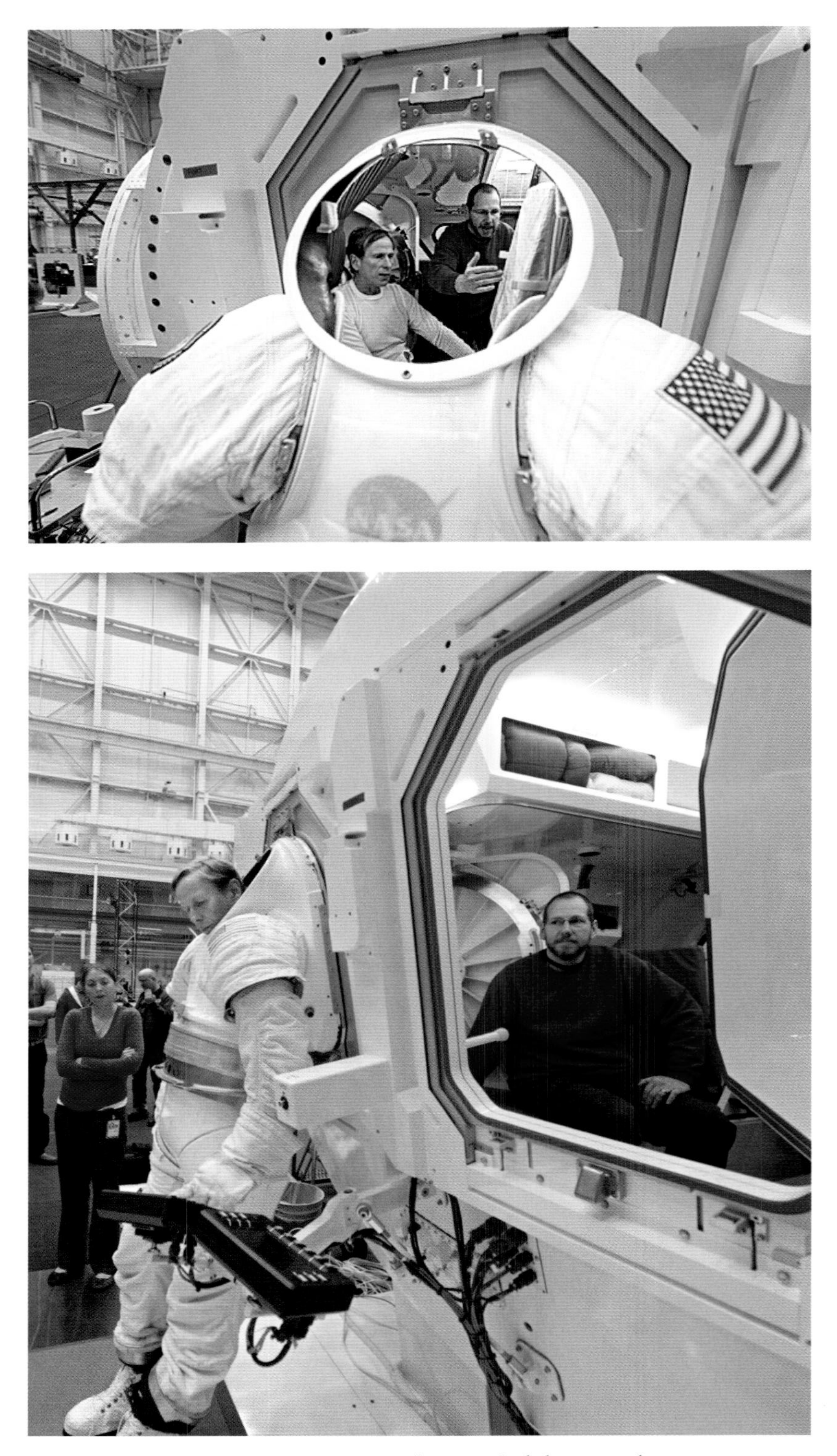

ASTRONAUT MIKE GERNHARDT *tests a "suit port" while exiting the cabin of the Multi-Mission Space Exploration Vehicle, as engineers evaluate the maneuver, which will be even more difficult in space.*

DIVERS PREPARE *to assist astronauts Akihiko Hoshide of the Japanese Aerospace Exploration Agency (JAXA) and Sunita Williams of NASA during their underwater training at NASA's Neutral Buoyancy Laboratory. The exercise is designed to mimic the pressurization and weightlessness of space.*

"AQUANAUTS" TRAIN *to use the DeepWorker 2000 submersible at the Aquarius habitat in Key Largo, Fla., as part of NASA's Extreme Environment Mission Operations (NEEMO) program, designed for studying human survival in space laboratories. Training underwater approximates environments of other worlds.*

ber has to be a jack of all trades," Whitson says, "because in a six-month mission, things are going to happen that we won't have predicted or just won't have time to train for."

The ISS doubles as a perfect training ground for the transit phase of any mission headed beyond Earth orbit, but NASA's future astronauts will need additional skills once they reach their destination. That's why the space agency recently had four people living 19 m (63 ft.) beneath the surface of the Atlantic for 12 days, using the ocean's buoyancy to simulate the low-gravity environment of an asteroid. In this case the "asteroid" is a 500-foot undersea landscape housing the National Oceanic and Atmospheric Administration's Aquarius Reef Base, a schoolbus-size undersea habitat similar to the type of habitats being developed for use in deep-space exploration missions.

It's all part of NASA's Extreme Environment Mission Operations (NEEMO). An annual test of the engineering challenges scientists expect to encounter on any extended journey from Earth, NEEMO doubles as a chance to identify the type of training future astronauts will need. "We're trying to get 15 or 20 years ahead of our program and understand the nuts and bolts of how we'll have our astronauts operate," says Mike Gernhardt, a NASA astronaut and NEEMO principal investigator.

This year's mission tested techniques for coping with a 50-second communications delay, the time it will take for radio signals to travel between a distant asteroid and Earth. Researchers also evaluated how best to distribute the four-person crew as they juggled tasks inside and outside the habitat, using exploration vehicles (small submersibles) and a simulated jet pack (an underwater thruster). Several new safety-restraint systems, designed to counter the fact that lunar astronauts spent nearly 3% of their time falling down, helped the "aquanauts" navigate the ocean floor. "Three years ago, no one had the first clue about how humans were going to operate on an asteroid," Gernhardt says. "Now we've got

THE CURRENT NEEMO TEAM *at the Aquarius habitat (from left): European astronaut Tim Peake; Cornell professor Steven Squyres; Aquarius Reef Base technicians James Talacek (in window) and Justin Brown; Japanese astronaut Kimiya Yui; NEEMO commander Dottie Metcalf-Lindenburger of NASA*

our arms around what it's going to take."

NEEMO's land-based equivalent is Desert RATS (Research and Technology Studies), an annual two-week program for field testing new equipment and procedures. "We needed to get the young people experiencing planetary exploration," says Joe Kosmo, a retired NASA engineer who started the program 15 years ago. Desert RATS has done exactly that, providing astronauts with hands-on time in a prototype rover dubbed the Space Exploration Vehicle (SEV). Unlike the old lunar rovers, the SEV has a pressurized cabin with "suit ports" containing spacesuits on the back end. To exit the vehicle, the astronauts enter the suits through hatches on the interior side. "Ten minutes later, your boot's on the surface," says Gernhardt, who is also the SEV mission manager. The SEV facilitates short walks and overnight journeys away from the lander, both of which are among the Desert RATS simulations.

Though astronauts currently use a Mark III spacesuit during Desert RATS, that will likely change, since future suits will need to be made with advanced composite materials. "You want something soft, comfortable and lightweight, with no protrusions or hard connection points and no bearings," says Kosmo, who retired in 2011 after overseeing NASA's spacesuit development from the Mercury program through the shuttle's conclusion. NASA's Whitson says the process is already under way. "Obviously we're going to have to change from where we are now," she says, "and folks are looking at lots of different options."

With so many changes in store, will astronaut allure go the way of pilot prestige? Most astronauts acknowledge the difference between the relatively routine task of shuttling to the ISS and the more daring mission of exploring the final frontier. "Just like airline pilot and test pilot, you're going to have [basic] astronaut and exploration astronaut," says Ferguson. Yet no matter what their role, the people with the right stuff to escape Earth's gravity will never be just ordinary folks.

CAPITALISTS OVER THE MOON

BY JEFFREY KLUGER

A handful of the world's most daring entrepreneurs are picking up where the space shuttle left off, betting big on transporting cargo and astronauts into space

If you're looking for a way to lose bil-lions of dollars, blow up a lot of expensive hardware and possibly even kill some people in the bargain, you can't pick a better field than space travel. There are reasons only governments have space programs: They usually have the cash to burn and the time to waste, and nobody asks them where the profits went. That, at least, is how things were—but it's not how they are anymore. In May 2012 the brand-new Dragon spacecraft—launched aboard the equally new

THE SPACEX DRAGON *(linked to the docking arm of the International Space Station) made history in May 2012 when it became the first commercial vehicle to deliver cargo in space.*

SPACEX CEO AND CHIEF DESIGNER ELON MUSK *(at Cape Canaveral launch site) stands beside the Falcon 9 rocket with its cluster of Merlin engines. Opposite, a simulated mission-control countdown (bottom), one of many that preceded the Falcon 9's dramatic liftoff (top) on May 22, 2012*

Falcon 9 rocket—splashed down in the Pacific Ocean after a nine-day mission during which it flew to orbit, rendezvoused with the International Space Station, exchanged cargo and reentered the atmosphere without incident.

It wasn't NASA that built the ship, nor was it Russia or China or the European Space Agency. It was Elon Musk, private citizen, owner of the California-based Space Exploration Technologies Corp. (SpaceX) and the current best bet to get Americans back into space in a big way—first to orbit and then maybe beyond.

Musk, 41, is a native South African and the inventor of what was once an obscure e-commerce service called PayPal. That little innovation made him a rich man, and in 2002, he rolled part of his fortune into SpaceX. The new company's goal was to reinvent the rocketry field by building simpler, more streamlined boosters as well as an Apollo-like spacecraft that he dubbed Dragon (in homage to the magical Puff), which will eventually be able to support a crew of up to seven people.

In the past decade, the fever-dream of having your own little space program has not been uncommon—at least not if you're very, very rich. What buying a sports team once was to the .0001%, going to space has become for the likes of Paul Allen, Richard Branson and others. Never mind the giants of NASA's magnificent past—the Armstrongs, Shepards, Glenns, Krafts—men with jet fuel in their blood who, over just eight years in the 1960s, guided the U.S. from a standing start to the surface of the moon. The future of the American space program is increasingly in the hands of people who are in the game to haul cargo and transport people—and make money in the process.

If commercial space travel seems to be a wide-open field, it may be because NASA has at least partly quit the stage. In February 2003, the shuttle *Columbia* disintegrated during reentry, eliminating any doubt that the shuttles, while beautiful, were a fragile, rococo, temperamental mess. In early 2004, President George W. Bush announced that the remaining fleet would be retired by 2010 and that NASA would return to its Apollo roots, building old-style expendable spacecraft. But the new program would have to rely on the same obstreperous bunch in Congress for funding and on the same ossified NASA bureaucracy for implementation. The whole enterprise soon fell back into the familiar pattern of overpromising and under-delivering, too little funding and too much infighting.

By 2006 the space agency faced reality and announced that it was establishing the Commercial Orbital Transportation Services (COTS) program, under which it would let the private sector take over the business of making taxi runs to low Earth orbit, freeing NASA to focus on missions to the moon and beyond.

Numerous companies leaped at the news, and by 2008, NASA had selected two winners in the COTS competition: SpaceX, which would be awarded a $1.6 billion contract to make 12 cargo runs to the International Space Station (ISS) from 2012 to 2015, and Orbital Sciences, a Virginia-based company that would get essentially the same deal. NASA was happy to let both companies fly missions and service the stations—but each would first have to prove it could build and launch the promised machines.

Musk's successful flight dazzled even the hard-to-impress folks at NASA. Like any consumer, NASA wants to be sure that what it's buying is solid and reliable. But also like any consumer, it's not immune to the less practical attractions of a shiny new piece of hardware. "It has a new-car smell," said astronaut Don Pettit after Dragon docked with the ISS and he opened the hatch and peered inside. Not long after, that car got sold. "We became a customer today," announced Alan Lindenmoyer, manager of NASA's commercial crew and cargo program, after Dragon safely splashed down.

AMAZON.COM FOUNDER JEFF BEZOS *now has a loftier goal: to make routine space tourism a reality through his new company Blue Origin.*

The real key to the spacecraft's appeal is not the smell but the hardheaded engineering behind it. Musk has been called the Steve Jobs of rocketry, and while that's not a claim he makes himself, he clearly has a Jobs-like respect for simplicity. SpaceX builds three different rockets: the Falcon 1, with one engine known as the Merlin; the Falcon 9, with nine Merlins; and the still-in-development Falcon Heavy, which will have three side-by-side clusters of nine Merlins. That tall-grande-venti lineup is a lot easier to manage than NASA's hodgepodge of multiple boosters from multiple suppliers with the dinosaurically strange shuttle atop them all. "People ask how it's possible to be safer but also more cost effective," says Garrett Reisman, a former shuttle astronaut who now works for SpaceX. "It's possible because complexity is the enemy of both."

SpaceX simplifies things in a lot of other ways too. The engines have improved cooling systems that allow them to run at lower temperatures, which means they can be built of less exotic metals. The manned version of the Dragon will have an emergency system that makes it possible for the crew vehicle to pop free and fly away from a Falcon booster that's about to blow—but the escape rockets will be built around Dragon's bottom, a simpler arrangement than the rocket tower that was bolted atop the old Apollo craft to do the same job."The number of major events that have to go right for a crew to survive an emergency on the Dragon is about half as many as with the shuttle," says Reisman.

But Musk isn't the only player in the game. Orbital Sciences, SpaceX's main competitor, has its own test flight scheduled for later this year, and the company didn't earn its COTS contract by accident. It's been around longer than SpaceX and has already established its ability to launch satellites. A success or two for Orbital coupled with a setback or two for SpaceX could change things completely.

Musk's plans to launch a human crew in a Dragon spacecraft by 2014 may put his company in a different league from Orbital's at the moment, but that field is getting crowded too. The Commercial Crew Development program (CCDev) is the follow-on to COTS, and as its name implies, it is focused on the business of getting people into orbit along with mere cargo. No fewer than five other companies are part of this race—including Boeing, a long-time NASA stalwart, as well as newcomers Blue Origin of Kent, Wash.; Sierra Nevada Corp. of Sparks, Nev.; and Paragon Space Development of Tucson, Ariz. Some of the companies are vying simply to provide support components such as environmental control systems; others, like SpaceX and Sierra Nevada, want to build and fly the entire spacecraft. Either way, NASA—by its own design—will be on the sidelines.

Flights to low Earth orbit—where real work with trained astronauts will be conducted—is only part of the private space industry. The other part is space tourism. Untrained passengers have taken rides to space before, beginning with Sen. Jake Garn, the onetime chairman of the Sen-

BLUE ORIGIN'S CONICAL SPACECRAFT GODDARD *(on its way back to its hangar in El Paso), was launched and landed on Nov. 13, 2006. The first stage in Blue Origin's New Shepard program, the Goddard is a vertical-takeoff, vertical-landing vehicle designed to transport tourists to space.*

TSC
THE SPACESHIP COMPANY
GALACTIC
VIRGIN GA

VIRGIN GALACTIC FOUNDER *Sir Richard Branson and Burt Rutan of Scaled Composites (above) intend to create a fleet of suborbital tourism vehicles. Their concept calls for the* SpaceShipTwo *to launch from the carrier aircraft* WhiteKnightTwo, *a technique that provides greater speed and lift than launching from the ground. A portion of Virgin Galactic's team posed (opposite) with the tandem vehicles in 2011 to celebrate the opening of their new hangar facility in Mojave, Calif.*

ate committee that oversaw NASA's budget, who once said that if he didn't get to fly on the shuttle he would not appropriate the space agency "another cent." Garn called that a joke, but NASA took it for something more and granted Garn a seat on the shuttle *Discovery* in 1985. The space senator spent a fair share of his time aloft doing little more than throwing up—suffering nausea so severe that NASA wags took to using his name as a new metric: The highest possible degree of spacesickness would forevermore be known as "one Garn."

Such possible unpleasantness aside, a number of private companies have gone into the business of building their own spacecraft designed to take wealthy vacationers on 17-minute suborbital jaunts at prices averaging about $200,000 per seat. The most high-profile of the cosmic capitalists is Sir Richard Branson, head of Virgin Airlines, whose new division Virgin Galactic has set up shop in the Mojave desert, with a sparkling new spaceport and a sparkling new vehicle—*SpaceShipTwo*—set to fly by the end of 2013.

Branson's idea isn't crazy—partly because he's partnering with Burt Rutan, the designer who launched two such manned missions to suborbital space within a five-week period in 2004. Rutan didn't use a traditional vertical rocket to get his astronauts off the ground but rather

FILM DIRECTOR JAMES CAMERON *(left) lends his name and support to Planetary Resources Inc., which—like his blockbuster* Avatar*—focuses on mining asteroids for water and precious minerals. The Leo telescope (rendered center) would analyze asteroids near Earth to determine their properties and commercial value. Based on the results, the robotic mining vehicle (top) would be employed for the extraction.*

an airplane-like mother ship, which carried the *SpaceShipOne* to a drop-off point of 50,000 ft. (15.2 km). At that altitude, the smaller ship flew free and lit its own rocket engines, which kicked it the rest of the way to the lowest boundary of space (about 62 mi., or 100 km). Branson's commercial version will rely on the same idea and the same general technology—albeit with a roomier, more passenger-friendly spacecraft.

So will it work? Well, the 500 folks (including such celebs as Ashton Kutcher, Angelina Jolie and Katy Perry) who have plunked down deposits certainly think it will. And the FAA, which just granted Branson permission to conduct the key flight tests necessary to certify the ship fit for use, is willing to give him a shot. But the reality might not match the dream.

For one thing, space travel will always be fearsomely dangerous, and no amount of liability waivers and flight insurance will change the fact that trying to get above the skin of the atmosphere carries a very real risk of catastrophe. The widely aired footage of the *SpaceShipOne*'s first flight made it painfully clear that the ship—and the people in it—were enduring a bone-rattling ride. The g-forces alone can be crushing, with passengers experiencing as much as six times the force of gravity on both ascent and descent. Yes, the five minutes of weightlessness in between will be fun, but the two days of flight orientation Virgin Galactic's customers will get before strapping in doesn't come close to the multiple years NASA astronauts spent preparing.

Branson is not alone in seeking passengers for his space seats. Blue Origin, which is owned by Amazon.com chief Jeff Bezos, also hopes to offer suborbital flights to paying customers, using an unconventional spacecraft that takes off like a vertical rocket and then settles back down the same way—landing on legs in much the same way as the old lunar modules. It looks cool, but it ain't easy—as McDonnell Douglas (now part of Boeing), in collaboration with Apollo astronaut Pete Conrad, discovered in the 1990s when it tested the Delta Clipper spacecraft, which was also designed for vertical takeoff and landing but never made it beyond test flights.

The most quixotic of all the private ventures—headed by the most quixotic of all visionaries—is Planetary Resources Inc., founded by X Prize creators Peter Diamandis and Eric Anderson and bankrolled by *Titanic* director James Cameron with Google billionaires Larry Page and Eric Schmidt. Planetary Resources' goal is not to transport tourists but to send robot craft out to mine asteroids for iron, nickel, platinum, gold, zinc and other substances in finite supply on Earth but plentiful in space. "Scarcity is contextual, and technology is an abundance-liberating force," said Anderson at the company's kickoff event in 2012. "We're going

SPACE REFINERY *Companies like Deep Space Industries hope to develop fuel processors like the one above, capable of extracting water and hydrocarbons from asteroids to power spaceships and craft like the Dragonfly (left) that can bag samples to bring back to Earth.*

to bring the solar system within our economic sphere of influence." Nice sentiment—but even enthusiasts concede that this could be a multicentury effort, requiring the development of technology capable of hauling space cargo exponentially heavier than anything we've remotely managed to move before.

But multicentury projects are nothing in the grand scope of space travel. Human beings were an entirely terrestrial species until the first lighter-than-air balloon carried a passenger aloft in 1783. We had never set a toe in space until Yuri Gagarin orbited Earth in 1961. A few hundred more years to master the new medium is, in relative terms, just a few minutes—time likely to pass even faster if the engine of private enterprise helps to carry the load.

THE 25 MOST INFLUENTIAL PEOPLE IN SPACE

BY DAVID BJERKLIE AND THE EDITORS OF TIME

As the frontiers of space expand, so do the opportunities for its explorers: to pilot spacecraft, spot planets, search for aliens—and share their passion. Here's an array of the most brilliant

Steve Squyres

MARS ROVER PACK LEADER

Earth's invasion of Mars (sci-fi writers had it backwards) began with planetary flybys in the 1960s; then came the Viking 1 and 2 landers in 1976 and the Pathfinder mission, with the first Mars rover, Sojourner, in 1997. Today, Cornell University astronomer Steven Squyres is spearheading a new scientific offensive as principal scientist of NASA's Mars Exploration Rover mission. Rover's robot geologists Spirit and Opportunity landed in January 2004 and have sent back more than 100,000 full-color images of Martian terrain as well as microscopic images and detailed analyses of rocks and soil surfaces. Squyres has also been an aquanaut at NASA's underwater lab in Florida, helping to plan for manned space missions in extreme environments.

Jacqueline Hewitt

DARK AGE ASTRONOMER

Science runs on curiosity and funding, and Jackie Hewitt, as director of MIT's Kavli Institute for Astrophysics and Space Science, has plenty of the former to attract the latter. Hewitt is a radio astronomer whose focus is on the cosmic Dark Age, the period between the Big Bang and the birth of the first stars and galaxies. The gestation of the early universe has so far been hidden from view, but radio astronomy aims to change that. Among the ambitious proposals Hewitt has championed is an immense farm of radio antennas to be deployed on the far side of the moon, well out of reach of any interfering static from Earth. While far out, such projects would be invaluable training grounds for the next generation of space scientists

Jerry Nelson

TELESCOPE INNOVATOR

"New telescopes and their instrumentation are at the heart of progress in astronomy," said the Kavli Foundation when it awarded its 2010 astrophysics prize to Jerry Nelson and two colleagues, Roger Angel and Ray Wilson. The great challenge in building large optical telescopes is to create a precise reflecting surface able to withstand distortion due to gravity, heat or cold. For decades, that meant a maximum mirror diameter of 6 m (19.7 ft.). Nelson, as technical leader for the twin 10 m (33 ft.) Keck telescopes on Mauna Kea, Hawaii, broke that barrier using segmented mirrors; Nelson is leading again in the design of the unprecedented Thirty-Meter Telescope, which will be the most advanced and powerful optical telescope on Earth.

R. Jay GaBany

STELLAR SHUTTERBUG

Gigantic observatories and orbiting space telescopes don't own astronomy. Not entirely. Smaller can sometimes be better, as astrophotographer R. Jay GaBany has proven. The galactic mergers that produce spiral galaxies like the Milky Way leave behind faint relics called stellar tidal streams. But because these cosmic fossils are so large, viewing them requires wider views and longer exposure times than are possible at major observatories. With a half-meter telescope located under the dark skies in the mountains of New Mexico, GaBany patiently produces unmatched images of stellar streams. Though officially an amateur, GaBany collaborates with the pros as a peer, earning scientific respect and accolades in the process.

Chryssa Kouveliotou

GAMMA-RAY ARGONAUT

As a child growing up in Greece, Chryssa Kouveliotou spent summer nights lying on the beach, searching the sky for falling stars and the tracings of satellites. Her determination to explore the heavens graduated to more exotic phenomena. "My first love was always gamma-ray bursts," recalls Kouveliotou, of NASA's Marshall Space Flight Center, "tremendous explosions that rock the universe like nothing else." One source of gamma rays is magnetars, the tiny, superdense remains of supernovas that generate the most powerful magnetic fields in the universe: Imagine a magnet strong enough to pull the keys out of your pocket from a distance halfway to the moon.

Andrea Ghez

BLACK-HOLE DETECTOR

Astronomers have long considered the Milky Way a mild-mannered galaxy. But when Andrea Ghez of the University of California, Los Angeles, mapped the galactic center with unprecedented resolution, she found stars moving at extraordinarily high speed, which meant they were orbiting something extraordinarily massive. (The key to the discovery was adaptive optics techniques, which compensate for the blurring effects of the atmosphere.) Her conclusion? Our milquetoast Milky Way has a monstrous black hole at its center, some 26,000 light-years from Earth. If a galaxy as sedate as our own harbors a massive black hole, such cosmic beasts may well lurk at the center of most galaxies.

Louis Allamandola

COSMIC CHEMIST

The building blocks of carbon-based life are found virtually everywhere in the universe. "Molecules from space helped to make the Earth the pleasant place that it is today," according to Louis Allamandola, founder of NASA's Ames Astrochemistry Laboratory. The puzzle of how these compounds form and how they combine with hydrogen, oxygen and nitrogen in the frigid, radiation-filled vacuum of space was solved by Allamandola and his colleagues by approximating those harsh conditions in the lab. Produced by dying, giant red stars, these carbon compounds have rained down on Earth since the origin of the solar system. As Allamandola has observed, "Even in death, the seeds of life are sown."

John Mather

OBSERVATIONAL COSMOLOGIST

The Big Bang left behind a telltale glow—still detectable today, nearly 14 billion years later—called the cosmic microwave background (CMB) radiation or, as cosmologist John Mather refers to it, "the accumulated trace of everything." Precise measurements of CMB are critical because any proposed model of the universe must be able to explain variations in it. Using data from the Cosmic Background Explorer satellite, Mather and colleague George Smoot made a map of the early universe, a work of cosmic cartography that won them Nobel Prizes in 2006.

Geoff Marcy

PLANET HUNTER

When astronomer Geoff Marcy decided to shift gears to look for planets in other star systems, it didn't seem like a brilliant career move at first. He teamed up with graduate student Paul Butler and, as Marcy remembers it, "When we told other astronomers about our search for extrasolar planets, they would often smile politely, look down at their shoes and change the subject." The duo's search took eleven years, but in 1995 they hit pay dirt. They quickly became the most prolific planet hunters and in 1999 were the first astronomers to find a multiple-planet system outside our own, where "three lovely planets" orbit a common star.

Liu Yang

CHINA'S FIRST WOMAN IN SPACE

Flight has a way of stirring ambitions. A school assembly inspired Liu Yang, who had wanted to be a bus conductor, to become a pilot. She joined her country's air force, racked up 1,680 hours as a fighter pilot and was recruited as a prospective taikonaut (as astronauts are called in China). Yang's historic mission into space in June 2012 was also China's most ambitious and complex; it included a manned space docking, a technically demanding procedure for both spacecraft and taikonauts but essential to the space station China wants to launch by 2020.

Jill Tarter

SEARCHER FOR EXTRATERRESTRIAL INTELLIGENCE

For three decades, Jill Tarter has been waiting for E.T. to call. She began her intergalactic watch in graduate school, and in 1984 she helped found the SETI Institute in Mountain View, Calif. Tarter also worked on developing a catalog of nearby stars that might have habitable planets. The main criteria? Stability for billions of years, enough time for intelligent life to evolve. In 2012 Tarter retired as director of SETI research to focus on fundraising for SETI's Allen Telescope Array. "The good news is that the tools we have now are getting better, faster and bigger," she told *New Scientist*. "We are finally acquiring a set of tools that is perhaps adequate for the task."

Michael Brown

PLUTO SLAYER

The summer of 2005 marked Pluto's last gasp. Caltech planetary astronomer Mike Brown and some colleagues were tracking objects on the fringes of the solar system, out where the comets roam, and they found one (later named Eris) that was larger than the ninth planet. In the 1990s, when such faraway objects first began to be spotted, the status of plutoids (initially called trans-Neptunian objects or dwarf planets) seemed to divide the world into Pluto huggers and haters. Brown's findings, however, finally tipped the balance toward Pluto's demotion. We may have lost a planet, but we've gained a more fascinating solar neighborhood.

oeb

DAWN EXPLORER

theory meets evidence that the rubber meets the road, in the view d astrophysicst Avi Loeb. Until now, our understanding has been eoretical about how the Dark Age of the early universe gave way to the awn of the first stars and galaxies. Loeb, in fact, was among the first to explore this frontier. But a new generation of telescopes promises ce a flood of data for theorists to consider. Of course, some work will speculative. The Milky Way is on a collision course with Andromeda and Loeb has run computer simulations of the outcome. In 5 billion ve'll know if he's right.

Elon Musk

ROCKET MAN

First he made a fortune with the Internet start-up PayPal greasing the wheels of e-commerce. Then he co-founded Tesla Motors, maker of the first electric sports car, and helped launch the alternative-energy provider SolarCity. But entrepreneur extraordinaire Elon Musk was just getting warmed up. Musk's company SpaceX made history in May 2012 as the first private company to deliver a payload to the International Space Station. SpaceX has a multiyear, dozen-mission contract with NASA, a growing list of international clients and plans to fly to the moon. Maybe Musk's ambition to send humans to explore Mars within the next two decades, a goal that struck skeptics as pie in the sky, isn't so crazy after all.

Sara Seager

EARTH-TWIN SEEKER

It has been a dramatic learning curve for planet hunters. First came the jaw-dropping experience of being able to detect such cosmically tiny objects so far away, then the excitement of finding hundreds of them. Today, astronomers like MIT's Sara Seager are sorting through these riches to find Earth twins. Seager's goal is to be able to recognize atmospheric gases, or biosignatures, that would signal life on a distant planet. And that, says Seager, would bring the Copernican Revolution full circle: Not only is Earth not the center of the universe; there are lots of other living planets out there as well.

Rashid Alievich Sunyaev

THEORY SURVIVOR

As a teenager growing up in the Soviet Union, Rashid Sunyaev loved to study history. But his father advised him not to pursue it as a profession. "He told me that he had several friends who were historians, and they were all shot or sent to prison," Sunyaev told an audience of students after he won the 2011 Kyoto Prize. So instead, he delved into the perturbations of the cosmos. Sunyaev, who divides his time between the Max Planck Institute for Astrophysics in Germany and the Space Research Institute in Moscow, has theorized about what variations in the cosmic microwave background radiation reveal about the structure of the universe and how matter behaves when spiraling into a black hole.

K. Radhakrishnan

INDIAN SPACE ORGANIZATION CHAIRMAN

Most people are surprised to learn that India launched a space program in 1962, the year John Glenn orbited the Earth. At the time, as even the father of the Indian space program, Vikram Sarabhai, acknowledged, there were critics who questioned "the relevance of space activities in a developing nation." Any remaining qualms were put to rest on Nov. 14, 2008, when India became the fourth nation to plant its flag on the moon with an unmanned lunar probe. A key person behind the Chandrayaan 1 mission was K. Radhakrishnan, now chairman of the Indian Space Research Organization. In 2013, India plans to send a rover to the moon as part of the Chandrayaan 2 mission.

Alan Guth

INFLATION CHAMPION

The Big Bang was born as an idea more than 50 years ago, but the theory had holes. For one, the universe couldn't have gotten so big so quickly. In 1980, Alan Guth devised a theory called inflation to fill the gap. Guth hypothesized an exponential expansion of space-time in nearly the first instant of the universe, during which it ballooned in size at least a trillion trillion times in less than a trillionth of a trillionth of a second. Talk about blowing up. If that isn't mind-boggling enough, such is the rarefied nature of the theory, notes Guth, that "it becomes very tempting to ask whether, in principle, it's possible to create a universe in the laboratory."

Carolyn Porco

SATURN FAMILY PHOTOGRAPHER

Black holes are not the only exotic game in town. As Carolyn Porco reminds us, mysteries abound in our own solar backyard. Porco is director of imaging for the Cassini-Huygens mission to Saturn. Since 2004, Cassini has discovered seven new moons as well as new rings. Hydrocarbon lakes have been spotted in the polar regions of the moon Titan and geyser-like plumes erupting from Enceladus. The combination of heat, liquid water and organic materials, says Porco, might just mean an environment hospitable to life. If life has arisen twice in our solar system, the odds suggest it has occurred a staggering number of times throughout the universe.

Adam Riess

COSMIC SPEED CLOCKER

Since the 1920s, astronomers have known the universe is expanding. But will the expansion go on steadily forever? Or will it eventually slow, stop, or even begin to contract? The answer is apparently none of the above. Johns Hopkins astronomer Adam Riess and others, using data from the Hubble telescope, have shown that the expansion is actually accelerating, a finding the Nobel Prize committee called "astounding." What's more, it appears that what is pressing the cosmic gas pedal is the mysterious entity "dark energy." Riess hopes Hubble's successor, the Webb Space Telescope, will help us understand why.

Brian Greene

SUPERSTAR STRING PLAYER

Science fans have never had it so good. As worthy heir to the cool-geek throne of Carl Sagan, Brian Greene packs a huge amount of smarts and enthusiasm into his passion for science. In addition to being director of Columbia's Institute for Strings, Cosmology, and Astroparticle Physics, Greene is co-founder of the World Science Festival and author of *New York Times* bestsellers *The Elegant Universe* and *The Fabric of the Cosmos* (both made into *NOVA* series on PBS) and other books. Greene doesn't just explain astrophysics; in his day job he ponders string theory, which aims to unify gravity and quantum physics.

Neil DeGrasse Tyson

HAYDEN PLANETARIUM DIRECTOR

When the American Museum of Natural History remodeled its Hayden Planetarium in the late 1990s, director Neil deGrasse Tyson quietly recast the solar system, and Pluto got the boot as a planet. A great fuss ensued, although astronomers would not officially demote Pluto until 2006. For Tyson, it was a perfect teaching moment—and as the host of several PBS *NOVA* series, including an upcoming remake of Carl Sagan's landmark series *Cosmos*, he knows teaching moments when he sees them. Tyson is also the author of 10 books, including a memoir, *The Sky is Not the Limit: Adventures of an Urban Astrophysicist.*

Martin Rees

ASTRONOMER ROYAL OF BRITAIN

In his post since 1995, Rees follows such celebrated predecessors as Edmund Halley, best known for predicting the return of the comet that came to bear his name. "Astronomers might seem the most helpless of all scientists," wrote Rees recently. "They can't do experiments on stars and galaxies, and human lives are far too short for us to watch most cosmic objects evolve." But fortunately, what the universe lacks in convenience, it makes up for in quantity. As a theoretical astrophysicist, Rees has pondered quasars, gamma-ray bursts, galactic nuclei and gravitational waves, as well as the possibility that our universe is merely one part of a vast multiverse.

David Spergel

DARKSIDE MAPPER

The satellite's name is a tongue twister—the Wilkinson Microwave Anisotropy Probe (WMAP)—but for astrophysicists like Princeton University's David Spergel, the data it has collected is pure gold. By revealing minute variations in the cosmic background radiation, WMAP has allowed researchers to fix the age of the universe at 13.75 billion years and conclude that the visible matter we take for reality makes up only a fraction (4.6%) of the universe; the rest comprises two little-known entities, dark matter and dark energy, providing a new frontier of epic mystery.

David Charbonneau

EXOPLANETOLOGIST

Imagine a character in a play stepping out of the shadows and into a spotlight at center stage. That's basically how Harvard University astronomer David Charbonneau finds planets orbiting other stars. When a faraway exoplanet passes in front of its star, it becomes visible, explains Charbonneau, and offers astronomers "unparalleled opportunities to determine the properties of the planet and its atmosphere." Charbonneau, who leads an exoplanet survey called the MEarth Project, began his search as a graduate student using a 4-in. telescope; today he uses the Kepler, Spitzer and Hubble space telescopes, as well as an eight-telescope array on Mount Hopkins in Arizona.

EARTH FROM SPACE

The images of the Earth's surface made possible today by the advanced technology on satellites and the International Space Station are stunning, sophisticated, and extremely useful

GLOBAL DARKNESS *A composite view of the world as it would appear if night reigned over the entire planet. It took the Suomi National Polar-Orbiting Partnership satellite 22 days, 312 orbits and 2.5 terabytes of data to create this image.*

DYNAMIC PLANET
Soaring, snow-capped peaks and ridges of the eastern Himalaya Mountains (far left) create an irregular white-on-red patchwork between major rivers in southwestern China. The image was taken by the Advanced Spaceborne Thermal Emission and Reflection Radiometer (ASTER) from NASA's Terra satellite. A vast alluvial fan (top) blooms across the desolate landscape between the mountain ranges that form the southern border of the Taklimakan Desert in China's XinJiang province. The river appears electric blue as it runs out of the mountains and then fans out into intricate, braided channels. Chile's Chaitén volcano (bottom) had been dormant for nearly 10,000 years when it erupted in 2008. For months the volcano remained active, and then, on Jan. 19, 2009, an explosive dome collapse occurred, and the volcano released a thick plume of ash and steam.

A DESERT THIRST *Over the past three decades, engineers in Saudi Arabia have drilled through sedimentary rock for a resource more precious than oil: "fossil" water that collected tens of thousands of years ago in aquifers buried far beneath the desert sands. Farmers are tapping these reserves to grow grains, fruits and vegetables in the desert. This series of images tracks agricultural operations in the Wadi As-Sirhan Basin as viewed by Landsat satellites in 1987, 1991, 2000 and 2012. New vegetation appears bright green while dry vegetation or fallow fields appear rust-colored. Dry, barren surfaces (mostly desert) are pink and yellow.*

SOUTHERN LIGHTS *Auroras, seen in northern and southern polar regions (the image here, captured on the International Space Station, is the southern aurora australis), occur when the Earth's magnetosphere is disrupted by a blast of hot ionized gas from the sun. Physics aside, as poet Robert Service wrote on seeing an aurora, "It is a throbbing, thrilling flame. . . . It swept the sky like a giant scythe, it quivered back to a wedge."*

AT THE EDGE OF OUR SOLAR SYSTEM

BY MICHAEL D. LEMONICK

The vast outer reaches of our planetary neighborhood, extending 3 billion to 5 billion miles from the sun, are filled with icy debris—and perhaps a few secrets

In the beginning, there was Pluto—the beginning being March 13, 1930, when a young astronomer named Clyde Tombaugh announced he'd found a telltale dot of light on a photographic plate at Lowell Observatory. Tombaugh was on the lookout for a new planet, and as he saw the dot inch its way across the sky from one night to the next, it was clear to him that this object was orbiting the sun just as a planet would. From that day onward, tiny Pluto, orbiting alone in the frigid nether regions of the solar system, was counted as a full-fledged planet, along with Earth, Mars, Jupiter and the rest.

It was, that is, until the summer of 2006, when the International Astronomical Union (IAU) downgraded poor Pluto, which at only about

KUIPER BOUND
The New Horizons spacecraft launched in 2008 will give astronomers their first close-up look at Pluto in 2015.

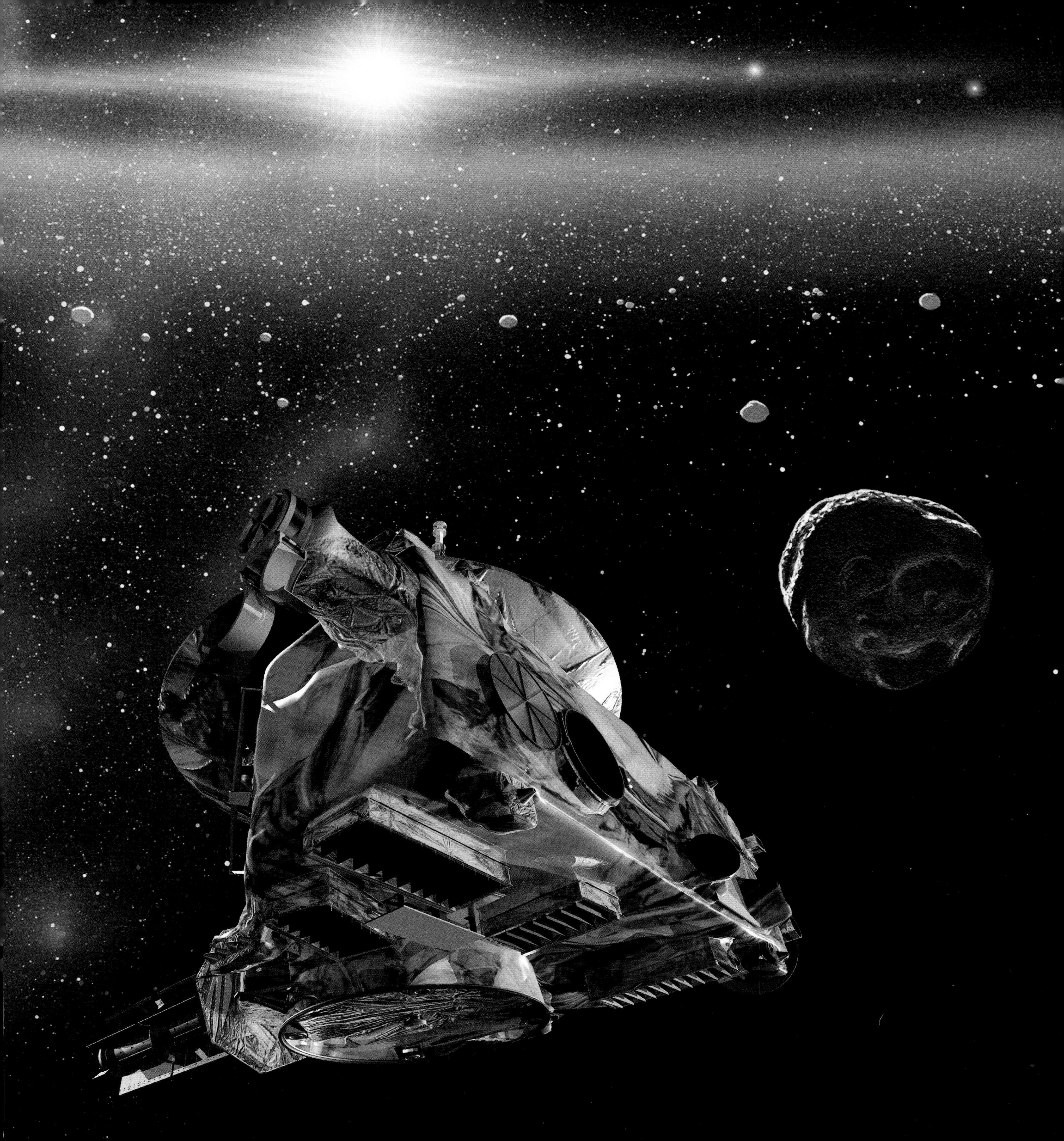

1,400 miles across (2,253 km) is significantly smaller than Earth's moon, to the second-rate status of dwarf planet. The decision caused much public outcry and triggered demands to reinstate the distant ball of ice to its former glory— but that just isn't going to happen. The best supporters can do at this point is join the "When I was your age, Pluto was a planet" group on Facebook.

The same set of discoveries that led to Pluto's demotion, however, has launched a revolution in astronomers' understanding of the outer solar system. Until the 1990s, everyone thought Pluto was literally the last frontier—the last world of any appreciable size that lay out beyond Neptune.

Since then, however, astronomers have realized that this icy world is just the most visible member of a huge population of similar objects, some of them nearly on a par with Pluto itself, in what's known as the Kuiper Belt. "For every asteroid in the asteroid belt, there are a thousand objects in the Kuiper Belt," says David Jewitt, the UCLA astronomer whose 1992 discovery of the first Kuiper Belt object (KBO) triggered the revolution. "It's staggering to me."

One of these worlds, called Eris, may even be bigger than Pluto. Another, Makemake, is only half Pluto's size but has oddly warm (though still relatively frigid) spots that might point to a fragmentary atmosphere. Then there's Quaoar and Sedna and Haumea and Orcus and more, all discovered over the past decade.

Even Pluto itself, dwarf planet though it may be, recently added a fifth moon to its own collection, which began with the 1978 discovery of Charon. By that count, Pluto now has more satellites in tow than Earth, Mars, Venus and Mercury combined. And in 2015, NASA's New Horizons space probe—launched in early 2006, when Pluto was still formally classified as a planet—will arrive for the first close-up look scientists have ever had at that frozen world. "New Horizons is now 80% of the way there," said lead scientist Alan Stern of the Southwest Research Institute in Boulder in early 2013. "We're on track for the encounter to begin in early 2015."

By the time that encounter ends a few months later, New Horizons, speeding past Pluto and its moons at more than 36,000 mph (57,936 km/h), will have sampled the dwarf planet's thin atmosphere, taken its temperature and snapped images that can pick out surface features as small as a football field.

It's important not only because a flyby of a tiny world more than 3 billion miles from Earth is such a technological tour de force. It's also the first time planetary scientists have had a good look at any Kuiper Belt object (KBO) in its natural state, a state that has been largely unchanged since the formation of the solar system some 4.5 billion years ago. "The Kuiper Belt is very, very cold," says Jewitt, with a temperature that hovers at about –380° Fahrenheit. "It's like a deep freeze."

Individual objects like Pluto, Eris and the thousand or so KBOs discovered so far haven't changed. But the Kuiper Belt as a whole has under-

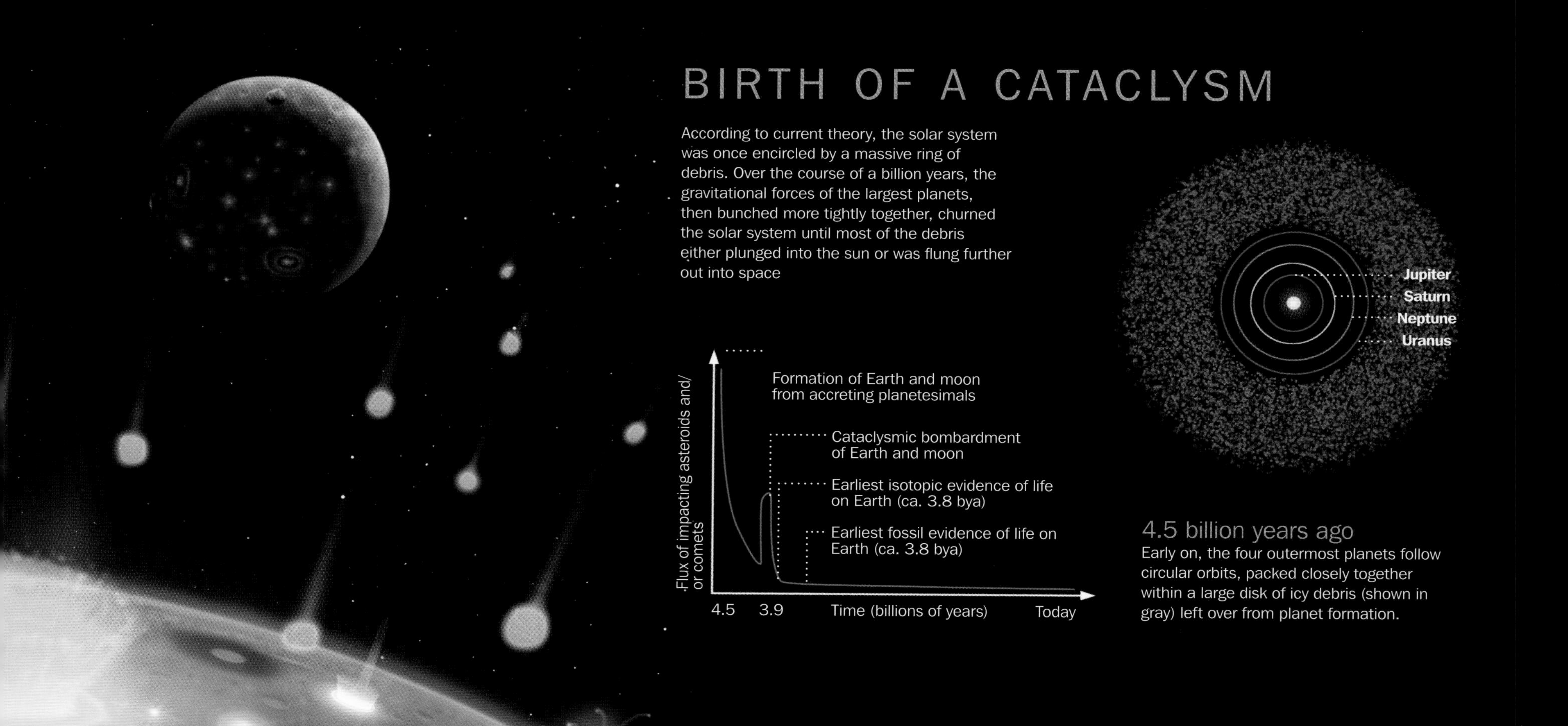

gone dramatic evolution over that time, according to the latest thinking. The chain of reasoning begins with a mystery: "We know there are vast numbers of objects out there," says Jewitt, "but if you add them all up, it amounts to about a tenth of an Earth-mass worth of material."

Finding lots more worlds the size of Pluto or even a little smaller might change that equation, but the odds appear to be against it. "If you'd asked me in back in 2005," says Mike Brown, the Caltech astronomer whose discovery of Eris in that year ended up toppling Pluto from the ranks of full-fledged planets, "I would have thought there were a lot more like it." But having looked long and hard over the past eight years or so, Brown says confidently, "There are no more bright KBOs to be found."

The problem is that such a meager amount of material, smoothly spread over the vast expanse of the outer solar system as it would have been in the early days, could never have clumped together to form anything as big as Pluto, Eris and the rest. "The consensus," says Jewitt, "is that there must have been a thousand times more to start with, and it somehow got whittled away and cleared out."

The "somehow" is a theory that raised plenty of eyebrows when it was first proposed a few years ago. Known as the Nice model, for the observatory on the French Riviera where it first came together, it posits that the four giant planets in the solar system were once more tightly bunched than they are today. They also had a brother—a fifth giant, of a size somewhere between Neptune and Jupiter.

Gravitational encounters with some of the larger Kuiper Belt objects moved all the giant planets except Jupiter gradually outward, until the pivotal moment when Saturn's orbit was exactly twice as long as Jupiter's, meaning that once every orbit, Jupiter and Saturn were in a straight line with the sun. Whenever that happened, their combined gravity—Saturn is 100 times as massive as Earth, and Jupiter three times as massive as Saturn—would have made the solar system act like a washing machine on spin cycle with a badly imbalanced load.

"When that happens," says Brown, "everything starts to bounce. Jupiter and Saturn shake around. Uranus and Neptune get flung outward . The whole solar system explodes dynamically." Jupiter and Saturn eventually drifted out of this disruptive configuration, but while it lasted, the beating would have been severe enough to throw the fifth giant planet out of the system entirely. Assuming it really existed, this long-lost world is now wandering alone in interstellar space. The Kuiper Belt would also have been devastated, with 99% of its mass either sent out into deep space as well, or sent plunging toward the sun in a rain of comets so thick it beggars the imagination.

Most of the craters on the moon came from just such an intense rain known as the Late Heavy Bombardment about 4 billion years ago, planetary scientists believe. And the Nice model is a plausible explanation for why it happened. "There's really been a lot of excitement about it and much discussion," says Jewitt. "It's totally changed how people think

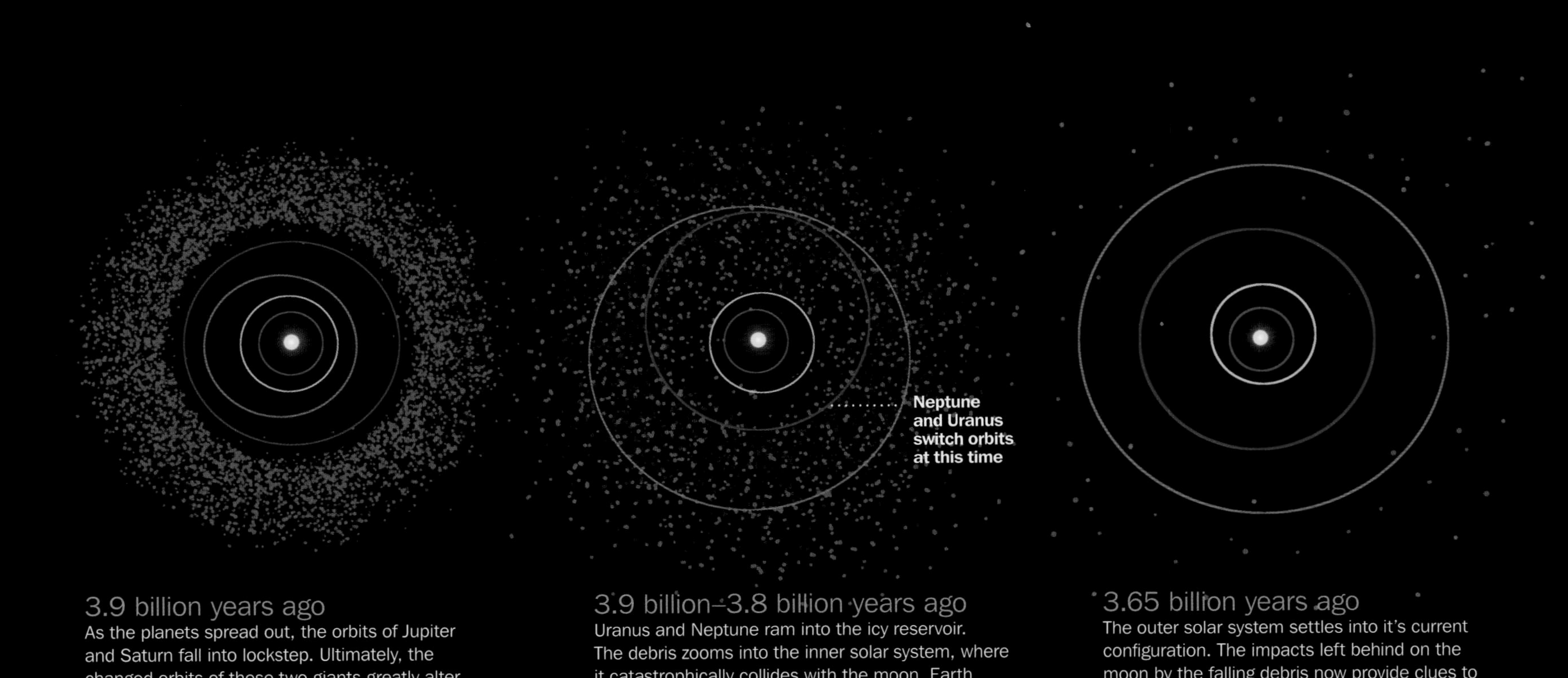

3.9 billion years ago
As the planets spread out, the orbits of Jupiter and Saturn fall into lockstep. Ultimately, the changed orbits of these two giants greatly alter

3.9 billion–3.8 billion years ago
Uranus and Neptune ram into the icy reservoir. The debris zooms into the inner solar system, where it catastrophically collides with the moon, Earth

3.65 billion years ago
The outer solar system settles into it's current configuration. The impacts left behind on the moon by the falling debris now provide clues to

THE OUTER REACHES

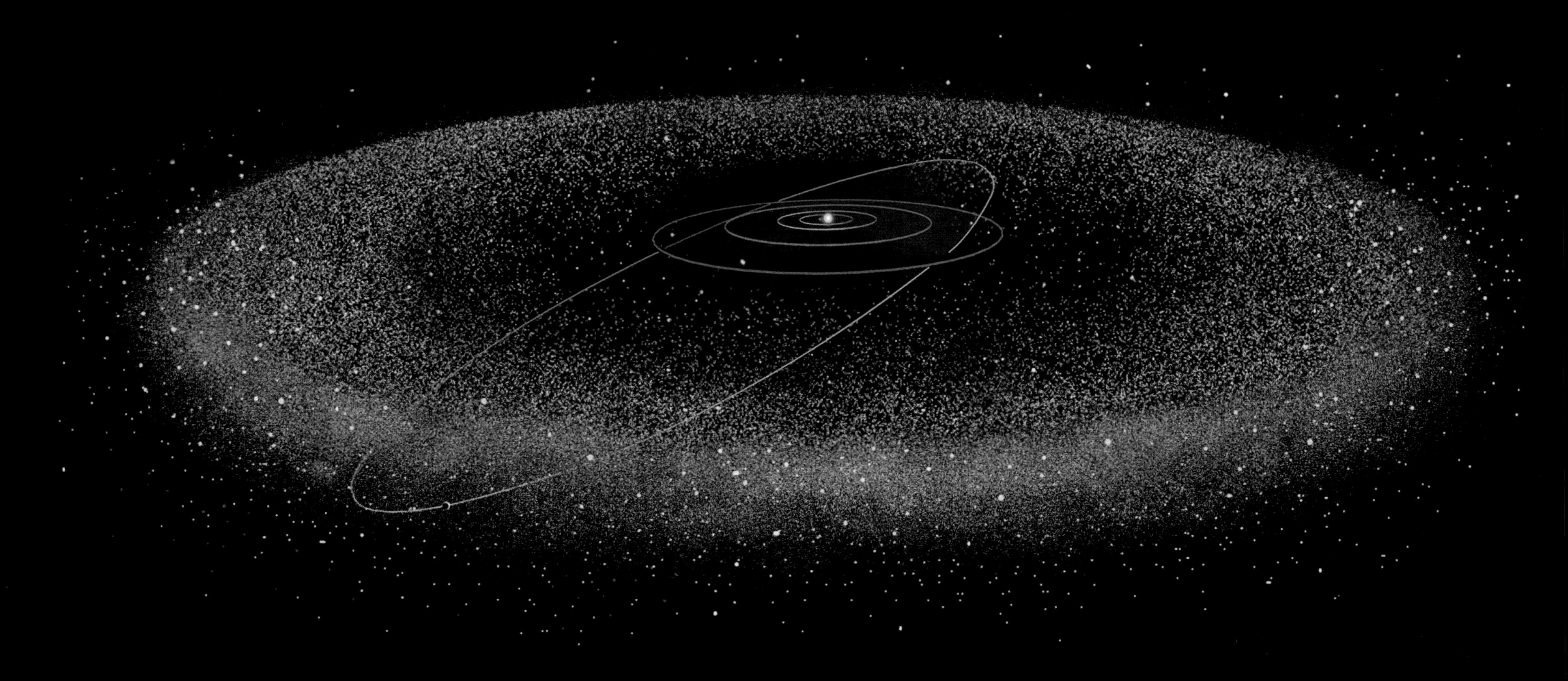

MEET THE NEIGHBORS

The same discoveries that led to Pluto's demotion have greatly added to our understanding of the solar system. Astronomers now know that Pluto is just the most visible member of a huge population of similar objects that reside in the Kuiper Belt; the largest discovered so far is Eris. Above, an artist's rendering of our expanded solar neighborhood; at right, a schematic showing the orbit of Eris and Pluto relative to the orbits of the other outer planets

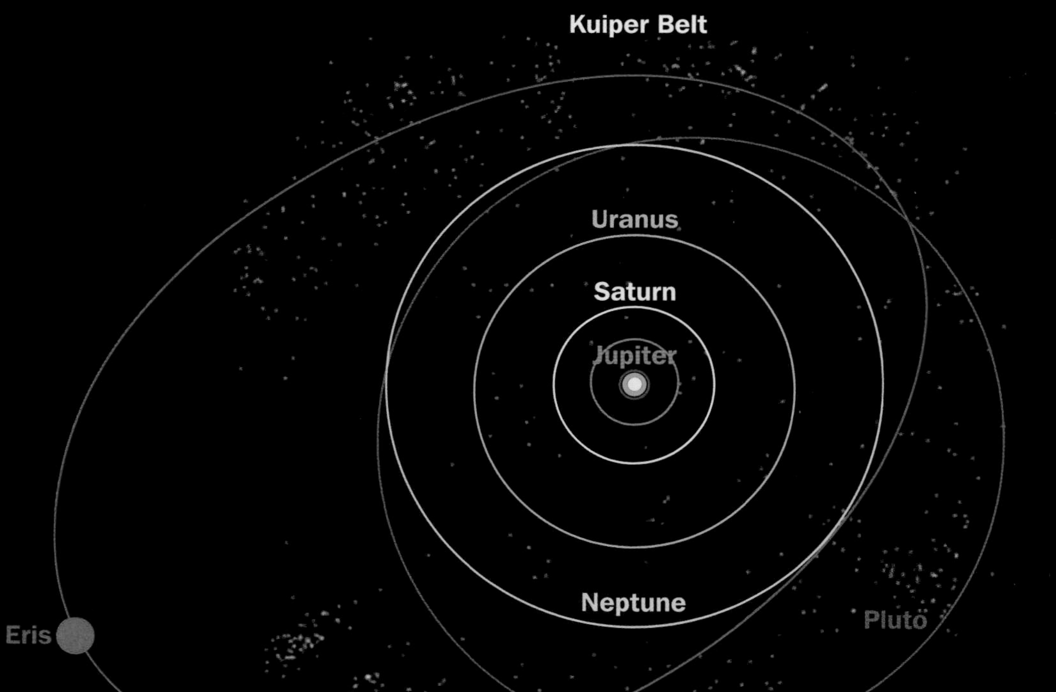

Outer Solar System

about the solar system. This Kuiper Belt business, which at some level is just finding a bunch of dots on the sky, has ended up affecting a lot of things."

The Nice model (or if you prefer, the unbalanced-spin-cycle model) comes from observations of the Kuiper Belt as a whole, but planetary scientists are eager to get a good look at its individual members as well. Using the Hubble Space Telescope, they've already had at least a blurry look at Pluto, which appears to have bright and dark splotches and a thin atmosphere made mostly of nitrogen. The atmosphere condenses into a planetwide frost when the tiny world's 246-year orbit takes it farthest from the sun.

Pluto's closest competitor in size, Eris, is about 9 billion miles from the sun, too far away for even Hubble to see it as more than a pinpoint of light. When Brown first discovered it in 2005, he made the reasonable assumption that its surface is similar to Pluto's. Using brightness as a benchmark, he gauged that Eris must be larger.

In 2011, however, Bruno Sicardy of the Paris Observatory managed to get an actual size measurement for Eris. He and his colleague watched as the KBO's orbit took it in front of a distant star, an event known as an occultation. Since Eris's orbital speed was by then well known, the timing of the star's disappearance behind the dwarf planet and subsequent reappearance allowed them to measure its size. Allowing for a margin of error, Eris turns out to be anywhere from a tiny bit smaller to a tiny bit bigger than Pluto.

Surprisingly, though, it turns out the assumption of a surface similar to Pluto's must be wrong: Eris, Sicardy told TIME, "is brighter than new-fallen snow." Eris also turns out to be much denser than Pluto: Mike Brown's 2005 discovery of Eris's moon Dysnomia allows scientists to calculate the parent body's gravity and mass. Combined with this new measurement of Eris's size, the density comes out at nearly 20% greater than Pluto's. Eris evidently has a bigger rocky core and a smaller icy coating than its rival.

The fact that the background star winked out abruptly and winked back in the same way suggests that Eris has no atmosphere (in Pluto's case, the dwarf planet's atmosphere makes occulted stars dim before winking). Another occultation, in 2012, allowed Spanish observers to conclude the same thing about Makemake, a roughly egg-shaped KBO that's half as large as Pluto and Eris.

Or at least, they decided there isn't a complete atmosphere. Because Makemake is so small and, at 4.6 billion miles away, so distant, the scientists knew even the slightest difference in viewing angle could mean the difference between seeing the star pass directly behind the KBO and seeing it barely kiss the edge. So they used seven telescopes in seven different locations—and some of them might, just might, have seen the background star dim before it disappeared. That could mean there are patches of atmosphere in areas where the planet is slightly warmer than average. It's also consistent with the fact that Hubble may have spotted some darker patches on Makemake's surface, that could in theory mark places that could absorb just a bit of extra sunlight and heat up.

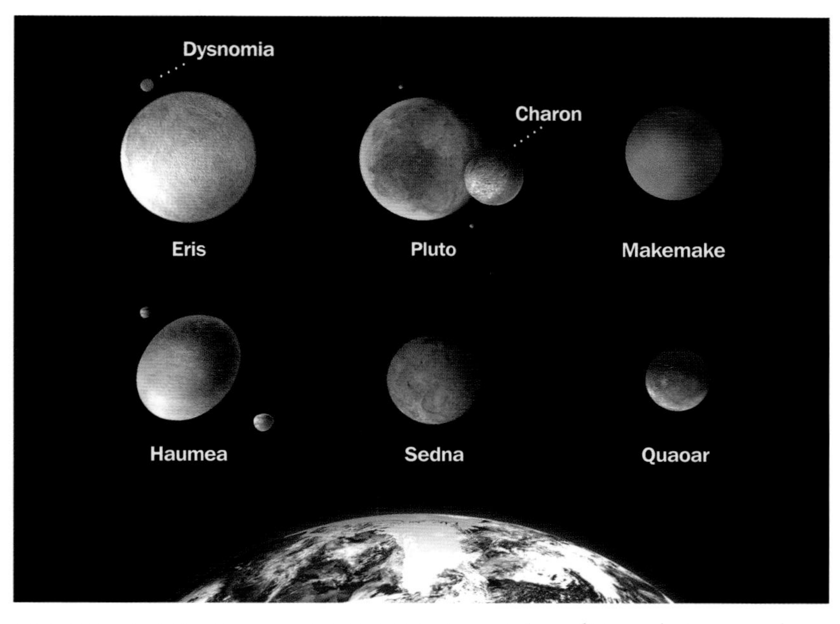

LARGEST KNOWN KUIPER BELT OBJECTS *An artist's rendition, relative to Earth. Top, from left: Eris and its moon Dysnomia; Pluto and the largest of its five moons, Charon; Makemake. Bottom: Haumea, Sedna and Quaoar*

But these observations are so difficult that nobody can say for sure what's really going on with Makemake. That's why New Horizons is going to be so crucial: It's probably the only chance scientists will get to look at a KBO in any detail at all (although the probe may be targeted at a second, if one is found along its path out of the solar system).

And while he's widely acknowledged as the scientist who toppled Pluto from its planetary pedestal—his 2010 popular book is titled *How I Killed Pluto and Why It Had It Coming*—Mike Brown is nearly as excited as anyone about the coming close encounter. "I tend to have a difficult time with space missions," he says, "because they look at just one object." It's only because they've catalogued hundreds of Kuiper Belt objects, after all, that planetary scientists can say anything meaningful about the solar system's frigid, outermost realm. On the other hand, he admits, "I'd rather see one up close than none."

Stern, meanwhile, rejects the idea that Brown or anyone else "killed" Pluto in the first place. *Real* planetary astronomers, he insists, still call Pluto a full-fledged planet, and they do the same with Eris and the other large KBOs, no matter what the International Astronomical Union might have ruled. "It's embarrassing when people who call themselves planetary scientists can't properly classify the objects that gave the field its name."

While the IAU may have its own definition of what's a planet and what isn't, says Stern: "I use the *Star Trek* test." When something comes up on the *Enterprise's* view screen, "the public knows in a millisecond if it's a star or a planet or a spaceship or whatever. This is common sense. "When we get to Pluto," he says defiantly, "we'll talk again."

NEW MYSTERIES OF THE COSMOS

BY MICHAEL D. LEMONICK

With powerful tools at their disposal, scientists are finding even more confounding puzzles, none more astonishing than the dark energy that propels the universe outward

You'd think the universe would have run out of surprises by now. Astronomers have been pointing their telescopes at the heavens ever since Galileo aimed his homemade spyglass upward in 1609. In the ensuing 400 years or so, those instruments have grown in size and sophistication far beyond anything the brilliant scientist could have imagined. Astronomers have used them to discover such astonishing things as supernovas, quasars, black holes, neutron stars and the Big Bang itself, which created an entire universe out of a seed far smaller than a subatomic particle nearly 14 billion years ago.

Yet despite all those remarkable discoveries, the cosmos is in many ways more mysterious now than it has ever been. Astronomers have

THE HUBBLE TELESCOPE'S *view of the Antennae galaxies colliding and birthing billions of stars*

known for nearly a century, for example, that the universe is peppered with galaxies—huge agglomerations of stars in blob shapes or pinwheel shapes, of which the Milky Way is just one of at least 100 billion. But they don't know how the galaxies formed or exactly when. They don't know why most galaxies, including ours, harbor gigantic black holes in their cores—collapsed objects so dense and with such powerful gravity that not even light can escape, each one millions of times as massive as a star—nor why some of those black holes erupt into the powerful cosmic flares known as quasars, nor how the black holes themselves formed.

The galaxies, stars and gas clouds that astronomers once believed to be the entire substance of the universe, moreover, turn out to be no more than 25% or so of the matter that's really out there. The other 75% is dark matter, made of something very different from ordinary atoms and molecules. "It's got to be something truly exotic," says Ohio State University astrophysicist David Weinberg, and nobody knows for sure what that might be.

Dark matter looks positively mundane compared with dark energy. When astronomers add up all the matter, light and dark, it still only accounts for no more than a quarter of the substance of the universe. The rest is something—nobody has even a plausible idea of what—that's forcing the already expanding universe to fly apart faster and faster as time goes on, as though some cosmic accelerator pedal is being stomped to the floor. "Anything we can say about dark energy at this point," asserts Princeton astrophysicist Michael Strauss, "is purely speculative."

Finally, there's the deepest question of all: What triggered the Big Bang itself, an event so paradoxical it seems like a Zen riddle? All of space and time were created in that single event. But if cause must come before effect and the Big Bang is the effect, how could there be a cause? What does the word "before" even mean in this context?

All these mysteries piled upon mysteries may sound so daunting that you wouldn't blame the scientists for throwing in the towel and taking up gardening. Instead, they're coming up with every trick they can think of to find out what's really going on in the universe.

Take dark energy, for example. It was first proposed by Albert Einstein nearly a century ago as a sort of accounting trick. His newly minted Theory of General Relativity dictated that the universe must either be expanding or contracting. As far as anyone could tell, it was doing neither, so Einstein invented the "cosmological constant," a kind of antigravity force that would perfectly balance gravity and permit a static universe. When Edwin Hubble found that the universe was expanding after all, Einstein discarded his invention and never looked back.

In 1998, though, two teams of astronomers independently discovered that the universe is expanding faster now than it was billions of years ago. There really is some sort of antigravity—or something that acts like it, anyway—turbocharging the cosmic expansion. When the evidence first started rolling in, Space Telescope Science Institute astronomer Adam Riess told

TIMELINE OF THE UNIVERSE

NASA's Wilkinson Microwave Anisotropy Probe (WMAP) was designed to measure the oldest light in the universe: cosmic microwave background radiation (CMB). Since its 2001 launch, WMAP has dutifully produced the first full-sky map of the microwave sky. Diagrammed here is the evolution of the universe over 13.7 billion years. Shortly after the Big Bang, the universe expanded exponentially, then slowed, then began expanding again.

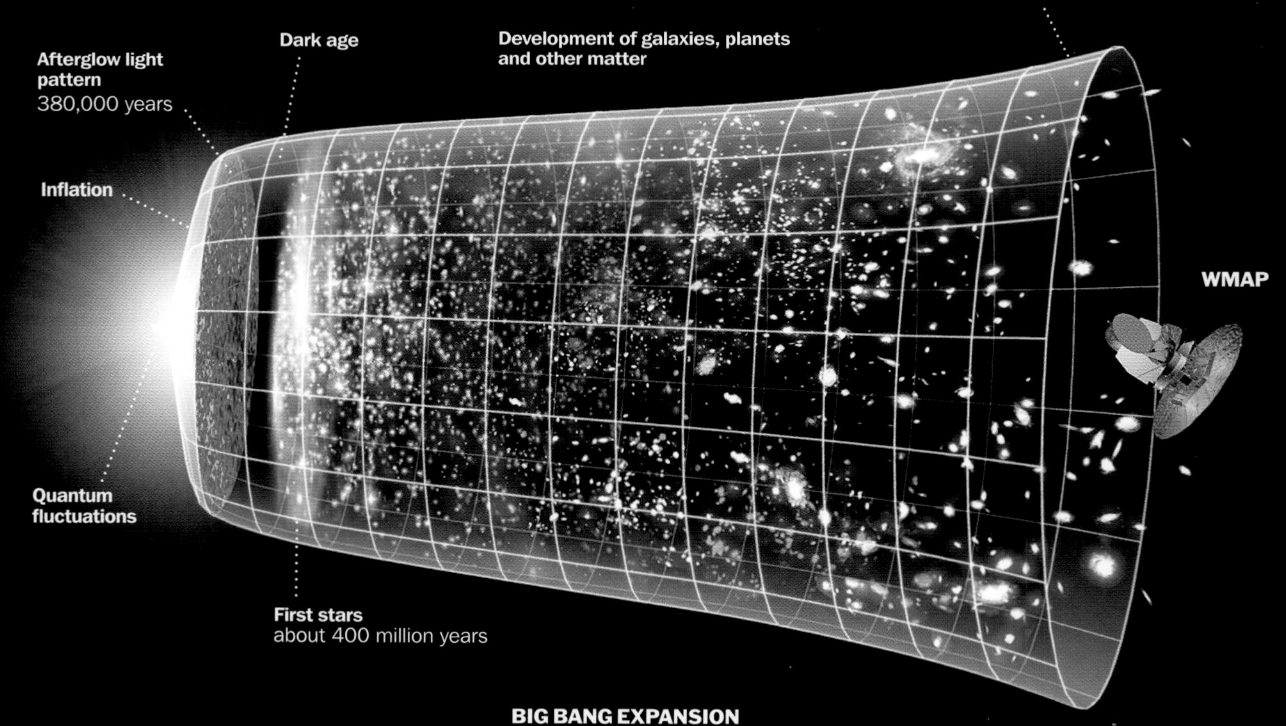

TIME in 2001, "I kept running the numbers through the computer, and the answers made no sense. I was sure there was a bug in the program." Last year, Riess shared the Nobel Prize for the discovery.

Both teams found dark energy by looking for distant supernovas, exploding stars so bright they can be seen from billions of light-years away. By comparing the supernovas' distances with how fast they're speeding away from Earth, the scientists could tell that something is making the universe expand faster.

The evidence is compelling, but dark energy is so weird that astronomers would love to confirm it with some sort of independent observation. That's just now becoming possible, thanks largely to the Baryon Oscillation Sky Survey (BOSS), part of the larger Sloan Digital Sky Survey. BOSS is based on the fact that for the first 400,000 years or so after the Big Bang, the entire universe was a hot, dense, primordial gas of subatomic particles that rippled and shuddered with density waves—precisely the way the air in a concert hall reverberates with sound waves.

When the universe cooled enough for the particles to condense into atoms and finally into stars and galaxies, those density waves were preserved, with more galaxies forming where the primordial gas was densest and fewer where it was relatively sparse. What BOSS does is look at the average spacing between dense clusters of galaxies. In the early days of the universe, that spacing should have been about 500,000 light-years, and without dark energy, the spacing should get steadily greater as the universe expands. Instead, says BOSS, the spacing gets wider at an accelerating rate. "It's a very powerful probe of dark energy," says Strauss. And it's consistent, he says, with what the supernova experts are seeing.

That's not nearly enough, however, to explain what dark energy actually is. To even begin to answer that question, you need telescopes far more powerful than the Sloan, whose 2.5-meter (8.2-ft.) diameter mirror, is "a wimp, by current standards," according to Strauss. Japan's 8.2-meter (26.2 ft.) Subaru telescope, in Hawaii, is now being fitted with a camera that will allow for a much deeper survey, and the Large Synoptic Survey telescope, due to see first light in 2019, will go even deeper. In that same year, the European Space Agency plans to launch its Euclid dark-energy space telescope, followed by NASA's Wide-Field Infrared Survey Telescope, or WFIRST—a device that suddenly looks to be far more powerful than expected because it may now incorporate an unused, Hubble-size telescope that the National Reconnaissance Office gave NASA in 2012.

The crucial question all these telescopes will try to answer: Does dark energy get steadily more powerful as the universe gets bigger, or does it vary up and down in strength? That could be a valuable clue to the nature of this bizarre and mysterious force that even Einstein abandoned as soon as he had an excuse to do so.

With dark matter, the picture is a little bit clearer. The evidence for its existence is indirect—without it, individual galaxies wouldn't have enough gravity to keep them from flying apart as they spin, and clusters

ALTHOUGH INVISIBLE, *dark matter can be measured by the way it distorts light from galaxy clusters like Abell 1689.*

RCW 86 *is what remains of the oldest recorded supernova—first observed by the Chinese in 185 A.D. This image was created by combining data from four different telescopes trained on the same point.*

of galaxies wouldn't have enough gravity to keep them from dispersing. But astronomers have a strong suspicion that dark matter probably consists of huge clouds of some sort of still-undiscovered subatomic particle. Theoretical physicists tell them that such particles plausibly exist, and if so, they should be detectable.

One possibility is that particles of dark matter could be produced by the fantastically powerful proton-on-proton collisions taking place at the world's biggest particle accelerator, the Large Hadron Collider, on the French-Swiss border, the same giant machine that in July 2012 discovered the long-sought Higgs boson, which gives other particles mass. Another, says Strauss, is direct detection. "If it fills the universe," he says, "it's right here. You can calculate how much there should be in this room." You can also calculate, based on a particle's presumed characteristics, how sensitive a detector you need to build to find it. Several such customized detectors are now operating in underground labs in Canada, Italy, Spain and the U.S. "Every few years," says Strauss, "people think they've found it, but it always goes away. One of these days, there will be a real we-found-it moment." If so, there will be Nobels all around.

All this attention to dark matter and dark energy is understandable, given that they make up fully 95% of the substance of the universe. But that hardly means there are no mysteries left in the other 5%. The birth of the very first stars in the universe, the first galaxies, the first black holes, all took place at more or less the same time.

It's not clear how these births were related to each other, however, so astronomers are pushing deeper and deeper into space and time to find out. Telescopes that see in ordinary, visible light aren't up to the job; over billions of years of cosmic expansion, starlight has been stretched so far that it has entered the realm of infrared, microwaves or even radio waves. So scientists use devices like the infrared-sensitive Spitzer Space Telescope. In 2012, it stared at two patches of sky for 400 hours each and managed to detect what might be the very first, ultramassive stars in the universe, burning furiously hot. Or they might be the very first black holes, burping superheated gas as they consumed everything within reach. "We know," says Weinberg, "that black holes a billion times as massive as the sun existed just a billion years after the Big Bang." To get so big when the cosmos was so young, they would have had to grow very rapidly, but nobody is quite sure how this might have happened. Astronomers may have to wait until 2018 for the James Webb Space Telescope, the Hubble's successor, to find out for sure.

They're on firmer ground in their understanding of quasars. When they were first discovered more than 60 years ago, these starlike objects that outshine an entire galaxy were utterly mysterious; since then, scientists have concluded that quasars are actually the hot exhaust spewed by black holes as they gobble up gas—the same phenomenon that Spitzer may have spotted but much more easily visible when the black holes are relatively nearby.

Every quasar is generated by a black hole, but not every black hole sports a quasar—at least not that we can see. In the modern universe, that's no surprise, says Strauss. "We're pretty sure every galaxy goes through a quasar phase," he explains, in which the galaxy's central black hole eats everything within reach and then exhausts the available supply of gas and dust. "Seeing quasars means we're seeing black holes grow."

The Milky Way's own black hole, as massive as 4 million stars put together, has long since eaten its fill and quieted down. The same goes for all modern galaxies. Still, says Strauss, the number of quasars in the early universe should match the number of black holes today—and the number of quasars seems far too low. The answer may be that early galaxies were very dusty, which might have shrouded the quasars from view. To penetrate the dust, astronomers are looking to the Atacama Large Millimeter Array (ALMA) in Chile, an unprecedentedly powerful 66-dish radio installation that can see even farther into the past than the Spitzer. And they're anxiously awaiting the even more powerful Square Kilometer Array (SKA), with thousands of dishes in South Africa and Australia.

SKA won't be online for several years, but in the meantime, there's no shortage of cosmic strangeness to explore. Theorists are still trying to understand exactly how stars explode into supernovas and why some supernovas leave behind superdense neutron stars while others leave black holes. They're just beginning to figure out how large galaxies, including the Milky Way, were assembled out of smaller galaxies over billions of years—and how the Milky Way is continuing to swallow small nearby galaxies whenever they come too close.

Observers, meanwhile, are still struggling to detect gravity waves, which Einstein predicted should sweep across space when black holes collide. The Laser Interferometer Gravitational Wave Observatory, based in Louisiana and Washington state, hasn't seen them so far in 10 years of looking. But indirect evidence is mounting that gravity waves do exist.

NASA'S SPITZER SPACE TELESCOPE *revealed patterns of light (from top, in infrared and enhanced views) from the first stars and galaxies within the universe.*

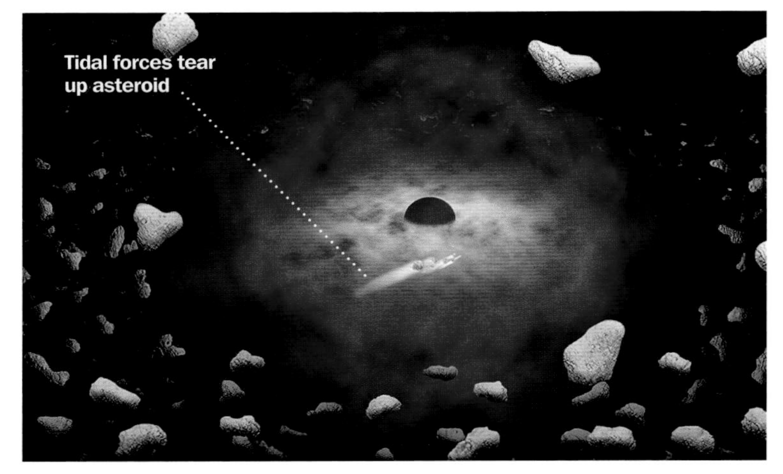

AN ARTISTS RENDERING *of the possible origins of X-ray flares that result when doomed asteroids are consumed by black holes*

THE SUPERMASSIVE BINARY BLACK HOLES *J0005-0006 and J0303-0019 are the oldest in the universe. At 13 billion years, they predate space dust and, as this rendering indicates, are surrounded only by gas rings.*

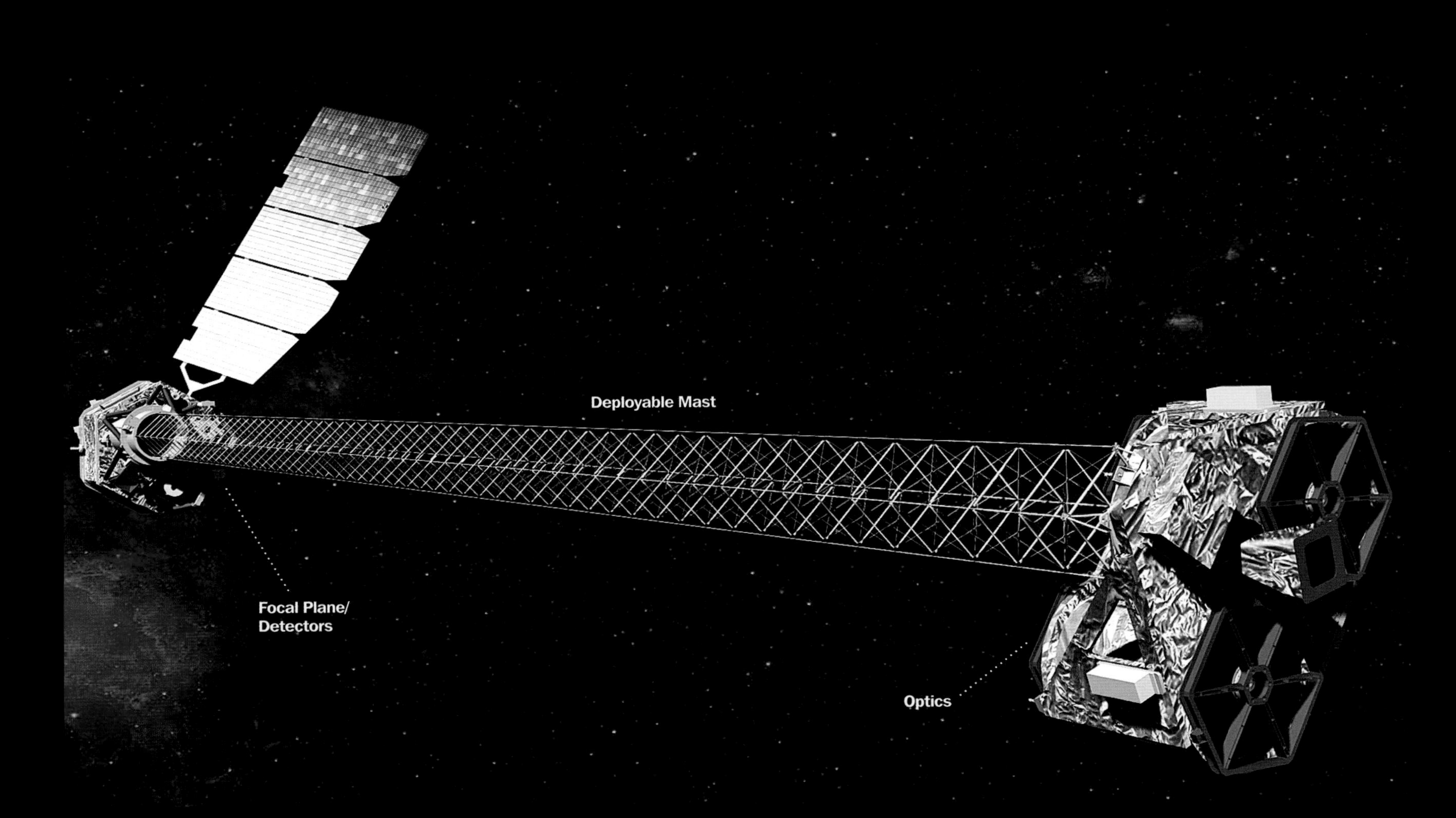

NUSTAR: THE NUCLEAR SPECTROSCOPIC TELESCOPE ARRAY

The Nuclear Spectroscopic Telescope Array *(NuSTAR, above)* was launched in June 2012 from Orbital Sciences Corp.'s Stargazer plane *(right)*. NuSTAR is an X-ray telescope designed to study black holes, particle acceleration and the remnants of supernovas. The image comparisons of galaxies beyond our own *(bottom right)* demonstrate NuSTAR's improved ability to capture high-energy X-ray images. At 100 times greater sensitivity and 10 times sharper resolution than previous equipment, it can help scientists probe deeper into the mysteries of the universe.

SPIRAL BEAUTY
Image of two black holes (magenta) in Caldwell 5 galaxy, taken by NuSTAR

The latest came in 2012 from NASA's orbiting Chandra X-ray Observatory. Astronomers have spotted what appears to be a supermassive black hole that has somehow been flung at several millions of miles per hour from the core of its home galaxy, which lies some 4 billion light-years from Earth. The most plausible explanation, improbable though it sounds, is that two huge black holes smashed together to form an even huger one. The collision sent gravity waves rippling into space, as Einstein said they should, but for some reason more went off in one direction—giving the black hole such a kick in the opposite direction that it will eventually escape to intergalactic space.

If that's what really happened, the implications are somewhat mind-bending. This kind of black-hole merger might be rare, but since there are at least 100 billion galaxies in the universe, it means gigantic black holes could well be roaming the dark places in between the galaxies, massive and utterly invisible.

Like most of what's going on out in the depths of the universe, they would pose no danger to earthlings, except perhaps to any sense of complacency on the part of astrophysicists: For no matter how many deep-space mysteries they solve, the cosmos always seems to come up with something completely new—and utterly unexpected.

LOOKING BACK ON LOOKING AHEAD

BY DAVID BJERKLIE

Our dreams of space travel took wing thanks to writers and artists centuries before engineers had the moxie and means to make those visions actually fly

Sorry NASA, but it did *not* begin with Sputnik in 1957. Never mind that the U.S. space agency officially celebrated the 50th anniversary of the Space Age in 2007. How quickly we forget the '30s, an era that was already mad about space travel—the 1630s, that is, a decade in which blockbuster space-travel fantasies were written by Johannes Kepler, Bishop John Wilkins, Francis Godwin and, a few years later, Cyrano de Bergerac. In fact, a case can be made that the first stirrings of the space age reach back nearly two millennia to the year 160 A.D., when Lucian of Samosata wrote what might be the first space opera (Lucian's tale purports to be a missing chapter from the *Odyssey*, in which a mighty whirlwind carries our hero's ship up and away to the moon and beyond).

A REACH TO EXCEED OUR GRASP
The telescope transformed heavenly bodies from abstract ideas into actual worlds we could dream of visiting (illustration from 1865 poem "The 'Monstre' Balloon").

Literary origins count, explains space artist and historian Ron Miller, author of *The Dream Machines*, "because astronautics is probably the only science that evolved from literature and art." And that's not just the "idea" of space travel. Literature and art pioneered many of the practical concepts that engineers would later try to execute. The internal-combustion engine, space suits, liquid-fuel rockets, solar sails, even nuclear propulsion, were all essentially invented by writers long before engineers got out their slide rules.

The influence went both ways, of course, says Miller. Writers like Edgar Allan Poe, who brought scientific verisimilitude to his comic space tale *The Unparalleled Adventures of Hans Pfall*, and Jules Verne, who understood the operation of rockets in a vacuum and was the first to appreciate the necessity of reaching escape velocity, were in thrall to the technological optimism of the 19th century. Between 1800 and 1865 (which marked the publication of Verne's *From the Earth to the Moon*), the public saw the introduction of ironclad warships, photography, electric lighting and motors, calculating machines, and transatlantic telegraph cables. Engineers were also building vast iron bridges, cutting canals through deserts and jungles, and spanning continents with railroads. "Engineers became heroes," says Miller. "People knew them by name, like they knew generals or kings or presidents."

The scientific realism of Poe and Verne set the standard for both the 19th and the 20th centuries. "A friend of mine who worked on the *Star Trek* series," says Miller, "told me that whenever they got to a place where they needed something technical but realistic, they would indicate on the story boards, 'insert science here.'" It was a model that seemed to have guided even the earliest motion pictures, including one made by Thomas Edison

TELLING IT LIKE IT WILL BE
In his 1865 classic and its 1870 sequel Round the Moon, *Jules Verne brought scientific verisimilitude to space fantasies.*

FLIGHTS OF FANTASY

1500
ROCKET CHAIR
Wan Hoo didn't reach the moon but did get a crater named for himself

1657
SOLAR ENERGY
Cyrano de Bergerac's fictional hero floated to the moon on the uplift of evaporating dew

1828
STEAM SPEED
Harnessing the power of steam propulsion seemed to be the answer to any transportation challenge

1835
FANTASY FUEL
The "Great Moon Hoax" sold newspapers and inspired illustrators to imagine fantastic voyages

MOON SHOT
*The Méliès brothers make a pioneering 1902 silent film, Le Voyage dans la Lune (*Voyage to the Moon*) based on Jules Verne's* From the Earth to the Moon *and H. G. Wells'* The First Men in the Moon.

in 1910 that chronicled a trip to Mars; by 1929, in Fritz Lang's movie Frau im Mond (*Woman in the Moon*), spaceships were impressively realistic. In the same year, Buck Rogers make his debut in American newspapers. He was followed in 1934 by Flash Gordon; portrayed by actor Buster Crabbe, he went on to an illustrious serial film career. Along for the ride were antigravity devices, magnetic waves, interplanetary electrical currents, and matter transmitters. But perhaps the Golden Age of the Spaceship, says Miller, was the period between the end of World War II and 1961, the year Yuri Gagarin became the first human in space. In movies, books and magazines, artists such as Chesley Bonestell (whose movie bona fides include *When Worlds Collide, Destination Moon, The War of the Worlds, Conquest of Space* and *Cat-Women of the Moon*) created the classic spacecraft that still inspire.

Landing on the moon did not end the fantasies. But the fantasies did not remain the same. Verne was confident that space travel was in our future. Kepler, too: In a remarkable letter to Galileo in 1610, Kepler was sure that, "Given ships or sails adapted to the breezes of heaven, there will be those who will not shrink from even that vast expanse." The motivation was undiluted romance. In the 20th century, however, the skies had darkened. Veteran astronaut and moonwalker John Young went so far as to say, "NASA is not about the 'Adventure of Human Space Exploration'; we are in the deadly serious business of saving the species." Young wasn't the only killjoy, even among the astronaut corps. Edgar Mitchell wrote, "Space exploration alone holds the promise of eventual escape from a dying planet, provided we wisely manage our resources in the meantime and actually survive that long." Even NASA director Mike Griffin stated in 2006, "In the long run, a single-planet species will not survive."

1893
NEW SHIPSHAPE
Space ships lost their fanciful sails and rigging as writers and artists created their own standards of realism.

1929
LUNAR MELODRAMA
Fritz Lang writes and directs the silent film classic Frau im Mond, or *Woman in the Moon*.

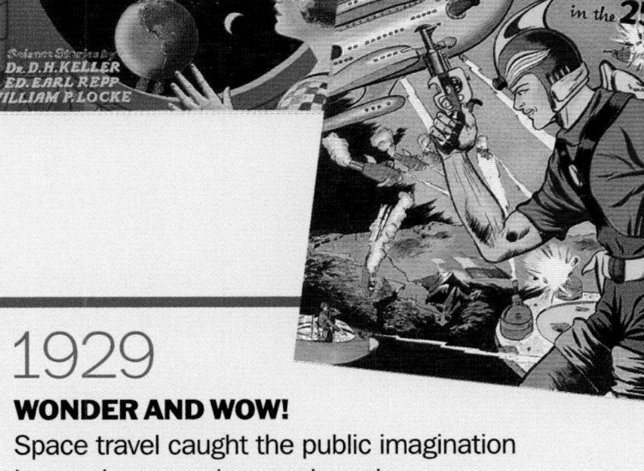

1929
WONDER AND WOW!
Space travel caught the public imagination in popular magazines and comics; Buck Rogers debuted as a newspaper comic strip, but also conquered comic books through the 1940s.

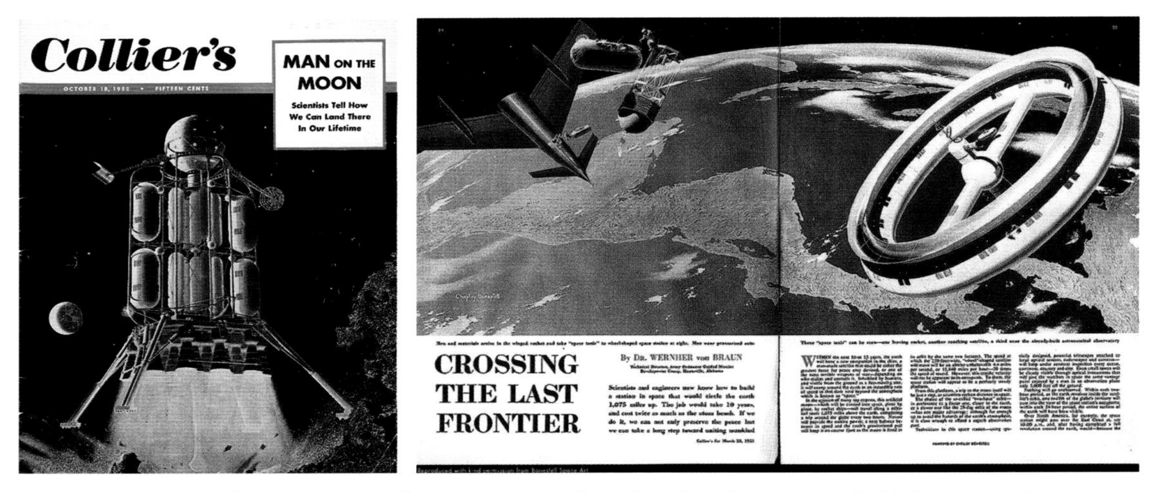

SPACE FUTURES *Rocket scientist Wernher von Braun made the leap from the Nazis to NASA; in 1952, with illustrators such as Chesley Bonestell, he helped Americans vault into space in the pages of* Collier's *magazine.*

The survivalist outlook was widely shared. Science fiction writers Robert Heinlein ("The Earth is just too small and fragile a basket for the human race to keep all its eggs in"), Ray Bradbury ("What's the use of looking at Mars through a telescope, sitting on panels, writing books, if it isn't to guarantee, not just the survival of mankind, but mankind surviving forever!") and Isaac Asimov ("Why not take what seems to me the only chance of escaping what is otherwise the sure destruction of all that humanity has struggled to achieve for 50,000 years?") also pushed the survival imperative. As did Carl Sagan and even black-hole theorist Stephen Hawking ("I don't think the human race will survive the next thousand years, unless we spread into space.").

Whatever the motivation, NASA, which is quick to point out that ideas of "warp drives" and "hyperspace" date back to the 1930s, has been dreaming big for quite some time. In the 1950s and '60s, Project Orion contemplated using nuclear-bomb powered propulsion (the project ended with the nuclear test ban treaty). And in 1994, the agency convened a brainstorming session to examine the emerging physics of faster-than-light travel, optimistically titled "Advanced Quantum/Relativity Theory Propulsion Workshop." The space folks examined theories of wormholes, tachyons, the Casimir effect, quantum paradoxes and the physics of additional space dimensions. While the participants conceded there were still holes in the case for wormholes, they did conclude that "there are enough unexplored paths to suggest future research." In other words, "Insert science here."

1931

BLAST OFF
Magazines tout liquid-fueled rocket pioneer Robert Goddard's vision

1936

FLASH FORWARD
First introduced as a comic strip, Flash Gordon starred in movies from 1936–41, and later, on TV too

1939

MISSION MOON
On the eve of World War II, dreams of heading to the moon

1955

SPACE ROUTINE
Space stations, shuttles and commuting are all in a day's work for Hollywood astronauts in *Voyage to the Moon*

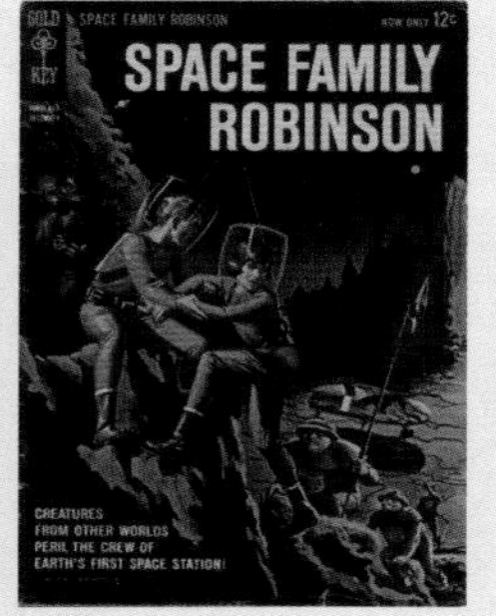

1962

BRING THE KIDS
Space families appeared in comic books and also on TV—in the 1965-68 series *Lost in Space*

DOUGHNUT DIGS *NASA hosted a design confab at Stanford University in the summer of 1975. One idea for a space colony capable of housing 10,000 to 140,000 inhabitants elaborated on von Braun's idea of using the doughnut shape called a torus.*

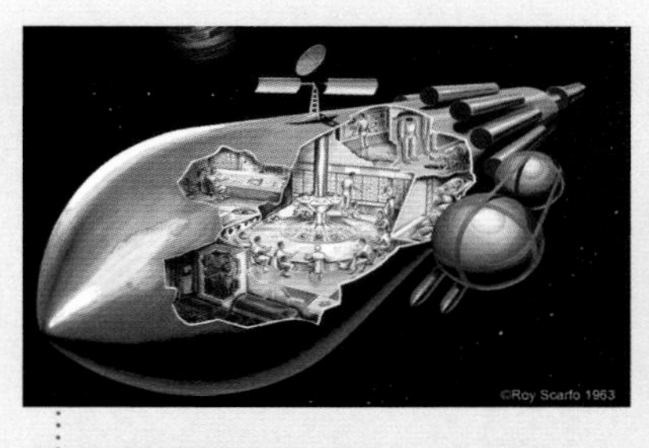

1963

NEXT STOP After nearby planets are colonized, ships will take folks to the outer planets as well

1964

FUTURAMA II General Motors updates its 1939 World's Fair vision to an even cooler space future for the 1964 World's Fair

1966

STARSHIPS FOREVER
The franchise begins with the original USS Enterprise in the *Star Trek* TV series

1977

SPACE OPERA The *Star Wars* epic that began a long time ago in a galaxy far, far away went on to world domination

THE FIRST MAN ON THE MOON

BY JEFFREY KLUGER

Human beings had long dreamed of walking on the moon. Then a quiet Midwesterner became the first man to do it. Neil Armstrong's life spanned just 82 years; his accomplishment is eternal

I DON'T KNOW HOW MANY TIMES AN AIRPLANE TRIED TO KILL Neil Armstrong, but with 78 combat missions in Korea and a test-piloting career that saw him at the stick of 900 flights, it's a fair bet he got used to the experience. I do know what was probably the last time he faced that kind of danger—just two years before his death in the summer of 2012—because I was along for the ride, and so were a couple of hundred other people.

It was in March 2010, and we were part of a large group returning from a six-country morale tour of military bases in the Middle East—a tour that also headlined Apollo veterans Gene Cernan and Jim Lovell. The trip included a flight over Iraqi airspace and down the Persian Gulf as well as a tailhook landing on the deck of a carrier and a catapult launch off the next day. The last leg of the journey, at least, would be routine: an overseas flight from Ramstein Air Base in Germany to Kennedy Airport in New York. Or at least it was supposed to be routine.

The weather forecast for much of the East Coast was brutal that day, and the reality was living up to those predictions, with lashing rains and powerful, almost lateral winds. The approach to the airport was murderous, with the plane lurching, the flight attendants urging calm. You could see the wings—to the extent they were visible through the soup outside the window—slightly but literally flapping. I had gotten lucky and snagged a seat in business class, but I could hear screaming coming from the rear of the plane, where the buffeting was worst. The astronauts were up in first.

As we made our final approach, there was a sudden gust of wind, the plane slewed to the side, and the pilot pulled up and broke off the landing. He climbed, circled back around and tried again. The result was the same—perhaps even worse. This time when he climbed, he gave up and headed off for a safer landing in Boston.

The scene in the cabin was what you'd expect, with loose belongings littering the aisles, air-sickness bags everywhere and the few children onboard wailing inconsolably. As soon as I could, I unbuckled and walked shakily up to the front. The first person I saw was Armstrong. He was sitting in his seat, his newspaper still open in front of him, serenely working on a Sudoku puzzle.

In some ways, one would expect nothing else. You don't pilot war planes, experimental jets and spacecraft without knowing how to retain your composure when things get dicey. But it was the depth of Armstrong's composure that defined the man. Lovell and Cernan were untroubled too, but like all the other passengers in what had been a marginally less storm-tossed first class, they were talking and even laughing with that high and happy release of tension that always follows danger. Armstrong was doing what Armstrong did best: remaining contained, controlled and wholly, utterly private.

That may have been part of his nature, but his experiences drew it out of him, too. Unlike most historical giants, Armstrong had his place in history bookmarked for him long before he got there. "The first man on the moon" had been a cultural construct for centuries—someone who would surely become real one day, but only when humanity had developed the improbable technology to make it so. No child ever dreamed of one day inventing the light bulb or explaining relativity—though when Edison and Einstein did so, they certainly joined the gallery of history's greats. But every child—particularly every boy—at some point angled for the gig Armstrong landed.

And yet when it finally happened, something had subtly changed. "ARMSTRONG TO BE FIRST MAN ON MOON," ran the headlines when NASA released its flight manifest for 1968 and 1969. Armstrong had worked hard and competed ferociously to land that spot, but ultimately, it was others who tapped him for fame.

That, of course, was the only way it could have been. The likes of Charles Lindbergh and the Wright brothers succeeded more or less under their own steam and on their own dime. Armstrong's mission required a global web of industrial, political and technological support—an invisible army of some 400,000 people. Astronauts like Armstrong simply walked point, which was fine; someone had to. But Armstrong also got to—had to—wave in parades, while the others went unthanked. That never ceased to trouble him. "He feels guilty that he got all the acclaim for an effort of tens of thousands of people," said his first wife, Janet, in James R. Hansen's authorized biography. "He's certainly led an interesting life. But he took it too seriously to heart."

During an earlier leg of our Middle East tour, I got a glimpse of Armstrong's deep discomfort with laurels he perceived were unearned. I asked him if I might get in touch with him after we got home to conduct an interview or two for a piece I was considering writing—one that would include some details about his past. "I confess that I haven't yet read your book," I said, referring to Hansen's *First Man: The Life of Neil A. Armstrong.*

"I don't have a book," Armstrong said. Snapped, actually. That was true; he hadn't written a book, and to say otherwise seemed to offend him on two levels. It would give him credit—again!—for work someone else had done, and it would imply that he was the kind of person who would write an autobiography in the first place. Both prospects were anathema to Armstrong—and he paid a personal price for that. When Hansen commented to Janet that Neil had won universal respect for never cashing in on his fame, her response was succinct: "Yes, but look what it's done to him inside."

There was a certain monastic power to Armstrong's famous reserve, on display many times in the pages of Hansen's book. It seemed that every time his moon-walking crewmate Buzz Aldrin appeared on a sitcom or an autograph show or at this or that glittering dinner, Armstrong would retreat a bit more, as if there were a fixed level of quiet grace his crew would be forever required to maintain. Whenever Aldrin tacked one way, he would have to tack the other.

That made it hard, at first, for me to understand why this most private of men had gone along on a meet-and-greet Middle East tour at all—and why, on two occasions afterward, he went back to Iraq and Afghanistan with Lovell and Cernan for more of the same glad-handing and public speaking he seemed to abhor. But for Armstrong, at least, this may have been a way of paying down a perceived debt: tithing his time to balance a moral ledger that only he was keeping.

And yet there were other, subtler debts he probably reckoned he would never quite be able to pay. During our visit to Ramstein, a very young boy—the son of a serviceman—was introduced to Gene Cernan. The father explained to his son that the man in front of him had walked on the moon—and had even lived there for a few days! The boy looked at Cernan with delight.

"Are you Neil Armstrong?" he asked.

Cernan, to his credit, laughed. "No, son. I'm Gene Cernan. But I went to the moon too."

The little boy turned to his father, puffed with indignation. "You said Neil Armstrong!" he protested.

Cernan, of course, was used to this. Lovell was used to it, Aldrin was used to it—the whole grand group of two dozen men who sailed moonward 40 years ago, did their surveying, took their pictures, collected their rocks, planted their flags and came safely and improbably home, were used to it. And Armstrong, surely, was more used to it than all of them. Others may have been able to carry that singular burden more easily and lightly than he did. But no one could have carried it with greater, more resolute strength. Armstrong had everything in his life a man could have wished for, except, perhaps, a sense of deep peace. If there is any consolation to be had from his passing, it is the hope that he has that now.

ON DESERT DUTY *In 2010, these former colleagues and lifelong mates engaged in a series of morale-boosting tours, one of which is remembered in this essay. Front row, from left, Cernan, Lovell and Armstrong pose with the troops at a U.S. military base in Kuwait.*

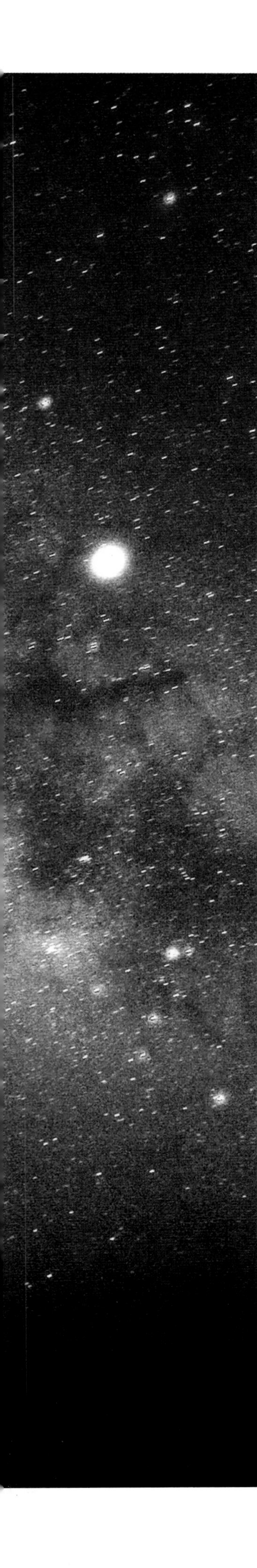

THE URGE TO EXPLORE

BY ANDREW CHAIKIN

Think space travel is over? The old methods may have gone stale, but the new dream of sustainable, affordable voyages beyond our world has only just begun to take off. And Mars is calling to us

IN 2000, WHEN I WAS THE EDITOR OF THE SHORT-LIVED magazine *Space Illustrated,* I had the privilege of working with the man I call the poet laureate of space exploration, Ray Bradbury. I commissioned him to write an essay about Mars and its enduring hold on our imaginations. What he gave me was not only the best evocation of the "why" of space I've ever heard, but a remarkably prescient description of where we are today as we struggle to become a spacefaring species. He called us "the Betweens" on a million-year journey from the cave to the stars, adding, "We must ignore the whispers from the cave that say, 'Stay.' We must listen to the stars, which say, 'Come.'" And he left no doubt it was a journey the universe requires us to make. "We have been given eyes to see what the light-year worlds cannot see of themselves," he wrote. "We have been given hands to touch the miraculous. We have been given hearts to know the incredible. Can we shrink back to bed in our funeral clothes? Mars says we cannot."

I thought about that when we lost Bradbury on June 5, 2012. Today, with the space shuttle no longer flying, many people have the mistaken belief that the space program is over. It isn't, of course, but until recently I often had the feeling it was in suspended animation. We have to look back 40 years, when Apollo 17's Gene Cernan and Jack Schmitt took the last footsteps on the moon, to see the most futuristic thing humans have ever done. As Cernan himself has said, it's as if President Kennedy grabbed a decade out of the 21st century and spliced it into the 1960s. Meanwhile, Congress and the White House spar over NASA's future, while American astronauts hitch rides on the Russian Soyuz until a replacement craft is ready to fly. The $100 billion International Space Station is finally complete but vastly underutilized, with only a fraction of its unique out-of-this-world research capabilities exploited by scientists. And NASA's spectacularly successful robotic explorers, which alone have continued to boldly go where no one has gone before, are threatened for years to come by draconian budget cuts—even though NASA's entire slice of the federal pie amounts to half a penny per dollar.

Still, every so often something happens to renew my hope. In the spring of 2012, it was the sight of SpaceX's Dragon capsule docking at the International Space Station on the first commercial resupply mission. Fifty-five years into the space age, the high cost of getting into low Earth orbit—still many thousands of dollars per pound—thwarts us like an overpriced toll bridge. SpaceX is one of a handful of companies striving, with NASA's help, to invent rockets and spacecraft that are not only reliable and safe but affordable, something NASA has been unable to achieve on its own. With a human-carrying version of Dragon in the works and a new effort to develop reusable boosters, I see at SpaceX the same passion and ingenuity that got us to the moon now aimed at a new "impossible" dream: making spaceflight sustainable.

But even that has been resisted by those who cling to the old way of doing things, the high-cost, government-pork space program that has been the rule since Apollo, a program that is as much about chasing jobs as about exploration. It seems we are "betweens," too, when it comes to the shifting paradigm of how to move beyond our home planet.

And yet, ironically, an antidote to our increasingly insular and polarized political climate lies beyond the atmosphere. One of the rewards of space exploration is that we are surprised at every turn—as we were by the discoveries of giant volcanoes on Mars, oceans under the icy crusts of the satellites of Jupiter and Saturn, and especially by the Hubble Space Telescope's revelation that a mysterious force that scientists call dark energy, acting counter to gravity, is speeding up the expansion of the universe. Exploration teaches us to leave our hubris at the door. And in a culture that shuns failure, it reaffirms that we must take risks in order to advance. As Kennedy said, we do these things "not because they are easy, but because they are hard."

So how do we find our way back to the endless frontier? We do what Ray Bradbury advised when he reminded us that every impossible dream that comes true begins with a leap of faith. "Jump off the cliff," he said, "and build your wings on the way down."

ROBONAUT 2 *(left) became the first humanoid robot in space in 2011. R2's mission is to help with cleanups and tasks requiring tools. Upgrades could eventually allow the robot to pitch in with astronauts on spacewalk chores.*

About the Authors

DAVID BJERKLIE
David Bjerklie is a former senior science reporter at TIME magazine, senior editor at TIME FOR KIDS and science writer/editor at TheVisualMD.com. He is also the author of children's books on butterflies, agriculture and environmental justice.

ANDREW CHAIKIN
Andrew Chaikin is a science journalist and space historian. He is the author of *A Man on the Moon: The Voyages of the Apollo Astronauts* (the main basis for the HBO miniseries *From the Earth to the Moon*); *Air and Space: The National Air and Space Museum Story of Flight*; *A Passion for Mars*; and *Mission Control, This is Apollo,* a book for young readers illustrated with paintings by Apollo moonwalker Alan Bean. Chaikin is also a commentator for National Public Radio's *Morning Edition.*

DANIEL CRAY
Daniel Cray, a contributor to TIME magazine since 1989, is the author of *Soaring Stones: A Kite-Powered Approach to Building Egypt's Pyramids*; *Friends from 4 a.m.,* a collection of short stories; and *The Reality Meltdown,* a novel.

JEFFREY KLUGER
A senior writer at TIME magazine, Jeffrey Kluger has written several books on space exploration, including *Lost Moon: The Perilous Voyage of Apollo 13* and *Journey Beyond Selene: Remarkable Expeditions Past Our Moon and to the Ends of the Solar System.* He is also the author of *Simplexity: Why Simple Things Become Complex (and How Complex Things Can Be Made Simple)* and *The Sibling Effect: What the Bonds Among Brothers and Sisters Reveal About Us.*

MICHAEL D. LEMONICK
Michael D. Lemonick is a senior science writer at the research organization Climate Central. He covered science for TIME magazine for more than 20 years and is the author of several books on astrophysics, including *Echo of the Big Bang*; *Other Worlds: The Search For Life in the Universe*; and the upcoming *Mirror Earth: The Search for Our Planet's Twin* (October 2012).

ALEX PERRY
Alex Perry, TIME's Africa bureau chief, has been a TIME correspondent since 2001, covering Asia, the Middle East and Africa via postings from Hong Kong, New Delhi and Cape Town. He has won numerous awards for his journalism, is presenter and co-creator of *Fishing for Trouble,* a documentary series exploring war zones through fishing, and the author of *Falling off the Edge: Globalization, World Peace and Other Lies,* and *Lifeblood: How to Change the World One Dead Mosquito at a Time.*

Credits

COVER Photo composite of Earth's moon and the Lagoon Nebula by R. Jay GaBany

ENDPAPERS Fornax Galaxy Cluster by NASA/JPL-Caltech/UCLA

FRONT MATTER **1** NASA/JPL-Caltech/Space Science Institute **3** NASA/ESA and the Hubble Heritage Team (STScI/AURA) **4** NASA/ESA and the Hubble Heritage Team (STScI/AURA) **7** NASA

STELLAR EXHIBIT **8** Joe McNally/Getty Images **10** NASA/JPL-Caltech/J. Hora (CfA) & W. Latter (NASA/Herschel) **12** NASA/JPL-Caltech/Univ. of Arizona **14** NASA/Goddard Space Flight Center/SDO

JUST AROUND THE CORNER **16** NASA/JPL/STSI **18** NASA/SDO **19** NASA/JPL-Caltech/Univ. of Arizona **20** Cassini Imaging Team, SSI, JPL, ESA, NASA **22** *(left)* NASA; *(right)* NASA/Zuber, M.T. *et al.*, Nature, 2012 **23** Courtesy *Science*/AAAS; *(inset)* NASA/Johns Hopkins University Applied Physics Laboratory/Carnegie Institution of Washington

ROCKING THE WORLD **24** Photo illustration by Lon Tweeten for TIME; asteroid by Stocktrek/Getty Images; Earth by World Persepctives/Getty Images **26** Chelyabinsk.ru/AP Images **27** The Asahi Shimbun via Getty Images

LIVE FROM MARS **28** NASA/JPL-Caltech/Malin Space Station Systems **30** Graphic by Lon Tweeten and Heather Jones for TIME **32** Brian van der Brug-Pool/Getty Images **33** NASA **35** C. Carreau/ESA

TRIUMPH OF THE PLANET HUNTERS **35** *(clockwise from top left)* Artist's Impressions by Tim Pyle/NASA; NASA/Ames/JPL-Caltech; C. Carreau/ESA; R. Hurt/NASA/JPL-Caltech **36** Paul Chinn/*San Francisco Chronicle*/Corbis **37** Carter Roberts/Eastbay Astronomical Society/NASA **38** *(top)* L. Calcada/ESO; Stephanie Mitchell/Harvard University **39** Graphic by Lon Tweeten **40** NASA/JPL-Caltech **41** *(left)* Courtesy of Caltech; G. Bacon/NASA/ESA

CHASING THEIR TAILS **42** Akira Fujii/David Malin Images

NEW EYES ON THE UNIVERSE **44** Dusko Despotovic/Corbis **46** *(top)* John Hill/Courtesy of LBTO.org; Joe McNally/Getty Images **47** Martin Bernetti/AFP/Getty Images(2) **48** *(top)* Joe McNally/Getty Images; Stephane Guisard/Courtesy of ESO **50** Steve Potter/Courtesy of SALT **51** *(top)* Courtesy of Webbtelescope.org; NASA

AFRICA'S EYE ON THE SKY **52** *(left)* Axel Mellinger/Central Michigan University; NASA **53** *(top to bottom)* SKA Africa(3); Popperfoto/Getty Images **54** Illustration by SKA Organization/TDP/DRAO/Swinburne Astronomy Productions **55** SKA Africa

E.T., ARE YOU CALLING US? **56** Courtesy of SETI Institute **58** Louie Psihoyos/Science Faction/Corbis **59** C. Rose/UC Berkeley **60** *(top)* NASA; design by Carl Sagan and Frank Drake/artwork by Linda Salzman Sagan/NASA **61** Laurie Hatch **62** Andy Kropa/Redux

ALIENS AMONG US **64** NASA/JSC/Michael Benson/Kinetikon Pictures

THE SEARCH FOR NASA'S MISSION **68** Artist's rendering courtesy of Sierra Nevada Corp. **70** Lockheed Martin/NASA **71** NASA **72** NASA **73** NASA

THEY'RE LAUNCHING ALL OVER THE WORLD **74** Imaginechina/Corbis **76** Li Gang/Xinhua Press/Corbis **77** *(top)* Kirill Kudryavtsev/AFP/Getty Images; Bill Ingalls/NASA//Corbis

A HALF-CENTURY IN SPACE **78-79** *(top, left to right)*; Keystone/Getty Images; MPI/Getty Images; SSPL/Getty Images; NASA(3); Time & Life Pictures/Getty Images; SSPL/Getty Images; Time & Life Pictures/Getty Images; NASA **78-79** *(bottom, left to right)* NASA (10) **80-81** *(top, left to right)* NASA (5); Newsmakers/Getty Images; AFP/Getty Images; Frederic Neema/Gamma-Rapho via Getty Images; Xu Haihan/ChinaFotoPress/Getty Images; Joe Raedle/Getty Images **80-81** *(bottom, left to right)* NASA/SpaceFrontiers/Getty Images; NASA(4); R. Kennicutt/JPL-Caltech/NASA; NASA(2); NASA/JPL-Caltech/UMD; NASA(2)

THE ASTRONAUTS OF THE FUTURE **82** Regan Geeseman/NASA Desert RATS **84** Bill Stafford/NASA(2) **85** NASA **86** Stephen Frink **87** Mark Widick/NASA

CAPITALISTS OVER THE MOON **88** NASA **90** SpaceX **91** SpaceX(2) **92** Joe Pugliese/Corbis Outline **93** Courtesy of Blue Origin **94** Mark Greenberg/Virgin Galactic **95** Gerald Holubowicz/Polaris **96** *(top)* Courtesy of Planetary Resources, Inc.(2); Desiree Navarro/Getty Images **97** Deep Space Industries(2)

THE 25 MOST INFLUENTIAL PEOPLE IN SPACE **98** Robert Barker/Cornell Univesity **99** Max Aguilera-Hellweg **100** *(left)* Michael Bolte/Courtesy of University of California, Santa Cruz; *(right)* Courtesy of R. Jay GaBany; *(bottom)* NASA **101** *(left)* Christopher Dibble; NASA **102** *(top)* Miguel Villagran/Getty Images; *(bottom)* ChinaFotoPress via Getty Images; *(right)* C. Rose/Courtesy of University of California, Berkeley **103** *(top)* Louie Psihoyos/Science Faction/Corbis; Stephanie Diani/*The New York Times*/Redux **104** *(left)* Max Aguilera-Hellweg ; Ramin Talaie/Corbis **105** *(top left)* Courtesy of MIT; *(top right)* ©Inamori Foundation 2011; *(bottom)* Pallava Bagla/Corbis **106** *(top to bottom)* Courtesy of MIT; Tania/A3/Contrasto/Redux; Courtesy of Johns Hopkins University **107** John Midgley/Courtesy of Brian Greene **108** Gabrielle Revere/Contour by Getty Images **109** *(top, left to right)* Susannah Ireland/Eyevine/Redux; John Jameson/Courtesy of Princeton University; *(bottom)* Stephanie Mitchell/Courtesy of Harvard University

EARTH FROM SPACE **110** NASA Earth Observatory image by Robert Simmon, using Suomi NPP VIIRS data provided courtesy of Chris Elvidge/NOAA National Geophysical Data Center **112** NASA Earth Observatory/USGS EROS Data Center **113** *(top)* NASA Earth Observatory/USGS EROS Data Center; NASA image created by Jesse Allen, using data provided courtesy of NASA/GSFC/METI/ERSDAC/JAROS and U.S./Japan ASTER Science Team **114-15** NASA Earth Observatory image created by Robert Simmon and Jesse Allen using Landsat data provided by the United States Geological Survey(4) **116** Image Science & Analysis Laboratory, NASA Johnson Space Center

AT THE EDGE OF OUR SOLAR SYSTEM **119** NASA/Johns Hopkins University Applied Physics Laboratory/Southwest Research Institute **120** Illustration by Mark A. Garlick/Space-art.co.uk; graph: D. Kring; simulations adapted from Gomes et al./*Nature* 2005; text from Science News **122** *(top)* Illustration by Don Dixon/Cosmographica; NASA **123** NASA

NEW MYSTERIES OF THE COSMOS **125** NASA/ESA and the Hubble Heritage Team (STScI/AURA)-ESA/Hubble Collaboration. Acknowledgement: B. Whitmore (Space Telescope Science Institute) and James Long (ESA/Hubble) **126** NASA/WMAP Science Team **127** NASA/JPL-Caltech/ESA/Institute of Astrophysics of Andalusia, University of Basque Country/JHU **128** NASA/JPL-Caltech/B. Williams (NCSU) **129** NASA/JPL-Caltech/GSFC **130** Illustrations by NASA/CXC/MIT/F. Baganoff/Chandra X-Ray Observatory Center **131** NASA/JPL-Caltech/R. Hurt (SSC) **132** *(top to bottom)* NASA/JPL-Caltech; Orbital Sciences Corporation; ESA/NASA/JPL-Caltech **133** NASA/JPL-Caltech/DSS

LOOKING BACK ON LOOKING AHEAD **135** Bridgeman Art Library **136** *(clockwise from top left)* no credit; Mary Evans Picture Library/Everett Collection; no credit(2); SSPL/Getty Images; MSFC/NASA(2) **137** *(clockwise from top left)* no credit(3); Everett Collection(2); Bridgeman Art Library **138** *(clockwise from top left)* Artwork by Chesley Bonestell/Reproduced Courtesy of Bonestell LLC(2); no credit; Everett Collection; Courtesy of Ron Miller from *The Dream Machines*; Silver Screen Collection/Getty Images; Courtesy of Ron Miller from *The Dream Machines* **139** *(clockwise from top left)* Don Davis/NASA Ames Research Center; Lucas Film Ltd./Twentieth Century Fox Film Corp./Photofest; no credit; GM Museum; Artwork by Roy Scarfo from *Beyond Tomorrow: The Next Fifty Years in Space* by Roy Scarfo with Dan Cole(2)

THE FIRST MAN ON THE MOON **140** Gamma-Keystone via Getty Images **141** Courtesy of Mike Whalen/Morale Entertainment Foundation

THE URGE TO EXPLORE **142** Baback Tafreshi/Photo Researchers/Getty Images

ABOUT THE AUTHORS **144** Robert Markowitz/NASA/JSC